Social Big Data Mining

Social Big Data Mining

Hiroshi Ishikawa
Dr. Sci., Prof.
Information and Communication Systems
Faculty of System Design
Tokyo Metropolitan University
Tokyo, Japan

CRC Press
Taylor & Francis Group
Boca Raton   London   New York

CRC Press is an imprint of the
Taylor & Francis Group, an **informa** business

A SCIENCE PUBLISHERS BOOK

CRC Press
Taylor & Francis Group
6000 Broken Sound Parkway NW, Suite 300
Boca Raton, FL 33487-2742

© 2015 by Taylor & Francis Group, LLC
CRC Press is an imprint of Taylor & Francis Group, an Informa business

No claim to original U.S. Government works

Printed on acid-free paper
Version Date: 20150209

International Standard Book Number-13: 978-1-4987-1093-0 (Hardback)

Visit the Taylor & Francis Web site at
http://www.taylorandfrancis.com

and the CRC Press Web site at
http://www.crcpress.com

Preface

In the present age, large amounts of data are produced continuously in science, on the internet, and in physical systems. Such data are collectively called data deluge. According to researches carried out by IDC, the size of data which are generated and reproduced all over the world every year is estimated to be 161 exa bytes. The total amount of data produced in 2011 exceeded 10 or more times the storage capacity of the storage media available in that year.

Experts in scientific and engineering fields produce a large amount of data by observing and analyzing the target phenomena. Even ordinary people voluntarily post a vast amount of data via various social media on the internet. Furthermore, people unconsciously produce data via various actions detected by physical systems in the real world. It is expected that such data can generate various values.

In the above-mentioned research report of IDC, data produced in science, the internet, and in physical systems are collectively called big data.

The features of big data can be summarized as follows:

- The quantity (Volume) of data is extraordinary, as the name denotes.
- The kinds (Variety) of data have expanded into unstructured texts, semi-structured data such as XML, and graphs (i.e., networks).
- As is often the case with Twitter and sensor data streams, the speed (Velocity) at which data are generated is very high.

Therefore, big data is often characterized as V^3 by taking the initial letters of these three terms Volume, Variety, and Velocity. Big data are expected to create not only knowledge in science but also derive values in various commercial ventures.

"Variety" implies that big data appear in a wide variety of applications. Big data inherently contain "vagueness" such as inconsistency and deficiency. Such vagueness must be resolved in order to obtain quality analysis results. Moreover, a recent survey done in Japan has made it clear that a lot of users have "vague" concerns as to the securities and mechanisms of big data applications. The resolution of such concerns is one of the keys to successful diffusion of big data applications. In this sense, V^4 should be used to characterise big data, instead of V^3.

Data analysts are also called data scientists. In the era of big data, data scientists are more and more in demand. The capabilities and expertise necessary for big data scientists include:

- Ability to construct a hypothesis
- Ability to verify a hypothesis
- Ability to mine social data as well as generic Web data
- Ability to process natural language information
- Ability to represent data and knowledge appropriately
- Ability to visualize data and results appropriately
- Ability to use GIS (geographical information systems)
- Knowledge about a wide variety of applications
- Knowledge about scalability
- Knowledge and follow ethics and laws about privacy and security
- Can use security systems
- Can communicate with customers

This book is not necessarily comprehensive according to the above criteria. Instead, from the viewpoint of social big data, this book focusses on the basic concepts and the related technologies as follows:

- Big data and social data
- The concept of a hypothesis
- Data mining for making a hypothesis
- Multivariate analysis for verifying the hypothesis
- Web mining and media mining
- Natural language processing
- Social big data applications
- Scalability

In short, featuring hypotheses, which are supposed to have an ever-increasingly important role in the era of social big data, this book explains the analytical techniques such as modeling, data mining, and multivariate analysis for social big data. It is different from other similar books in that it aims to present the overall picture of social big data from fundamental concepts to applications while standing on academic bases.

I hope that this book will be widely used by readers who are interested in social big data, including students, engineers, scientists, and other professionals. In addition, I would like to deeply thank my wife Tazuko, my children Takashi and Hitomi for their affectionate support.

Hiroshi Ishikawa

July, 2014   Kakio, Dijon and Bayonne

Contents

1

Social Media

Social media are indispensable elements of social big data applications. In this chapter, we will first classify social media into several categories and explain the features of each category in order to better understand what social media are. Then we will select important media categories from a viewpoint of analysis required for social big data applications, address representative social media included in each category, and describe the characteristics of the social media, focusing on the statistics, structures, and interactions of social media as well as the relationships with other similar social media.

1.1 What are Social Media?

Generally, a social media site consists of an information system as its platform and its users on the Web. The system enables the user to perform direct interactions with it. The user is identified by the system along with other users as well. Two or more users constitute explicit or implicit communities, that is, social networks. The user in social media is generally called an actor in the context of social network analysis. By participating in the social network as well as directly interacting with the system, the user can enjoy services provided by the social media site.

More specifically, social media can be classified into the following categories based on the service contents.

- *Blogging*: Services in this category enable the user to publish explanations, sentiments, evaluations, actions, and ideas about certain topics including personal or social events in a text in the style of a diary.
- *Micro blogging*: The user describes a certain topic frequently in shorter texts in micro blogging. For example, a tweet, an article of Twitter, consists of at most 140 characters.

- *SNS (Social Network Service)*: Services in this category literally support creating social networks among users.
- *Sharing service*: Services in this category enable the user to share movies, audios, photographs, and bookmarks.
- *Video communication*: The users can hold a meeting and chat with other users using live videos as services in this category.
- *Social search*: Services in this category enable the user to reflect the likings and opinions of current search results in the subsequent searches. Other services allow not only experts but also users to directly reply to queries.
- *Social news*: Through services in this category the user can contribute news as a primary source and can also re-post and evaluate favorite news items which have already been posted.
- *Social gaming*: Services in this category enable the user to play games with other users connected by SNS.
- *Crowd sourcing*: Through services in this category, the user can outsource a part or all of his work to outside users who are capable of doing the work.
- *Collaboration*: Services in this category support cooperative work among users and they enable the users to publish a result of the cooperative work.

1.2 Representative Social Media

In consideration of user volumes and the social impact of media in the present circumstances, micro blogging, SNS, movie sharing, photograph sharing, and collaboration are important categories of social big data applications, where social media data are analyzed and the results are utilized   as one of big data sources. The profiles (i.e., features) of representative social media in each category will be explained as well as generic Web, paying attention to the following aspects which are effective for analysis:

- Category and foundation
- Numbers
- Data structures
- Main interactions
- Comparison with similar media
- API

1.2.1 Twitter

(1) Category and foundation

Twitter [Twitter 2014] [Twitter-Wikipedia 2014] is one of the platform services for micro blogging founded by Jack Dorsey in 2005 (see Fig. 1.1).

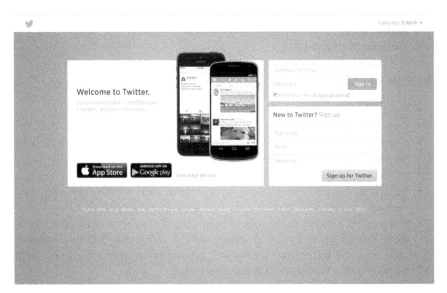

Figure 1.1 Twitter.

Color image of this figure appears in the color plate section at the end of the book.

Twitter started from the ideas about development of media which are highly live and suitable for communication among friends. It is said that it has attracted attention partly because its users have increased so rapidly. For example, in Japan, when the animation movie "Castle in the Sky" by Hayao Miyazaki was broadcast as a TV program in 2011, there were 25,088 tweets in one second, which made it the center of attention.

(2) Numbers

- Active users: 200 M (M: Million)
- The number of searches per day: 1.6 B (B: Billion)
- The number of tweets per day: 400 M

(3) Data structures

(Related to users)
- Account
- Profile

(Related to contents)
- Tweet

(Related to relationships)
- Links to Web sites, video, and photo
- The follower-followee relationship between users

- Memory of searches
- List of users
- Bookmark of tweets

(4) Main interactions

- Creation and deletion of an account.
- Creation and change of a profile.
- Contribution of a tweet: Tweets contributed by a user who are followed by another user appear in the time line of the follower.
- Deletion of a tweet.
- Search of tweets: Tweets can be searched with search terms or user names.
- Retweet: If a tweet is retweeted by a user, the tweet will appear in the time line of the follower. In other words, if the user follows another user and the latter user retweets a certain tweet, then the tweet will appear in the time line of the former user.
- Reply: If a user replies to a message by user who contributed the tweet, then the message will appear in the time line of another user who follows both of them.
- Sending a direct message: The user directly sends a message to its follower.
- Addition of location information to tweets.
- Inclusion of hash tags in a tweet: Tweets are searched with the character string starting with "#" as one of search terms. Hash tags often indicate certain topics or constitute coherent communities.
- Embedding URL of a Web page in a tweet.
- Embedding of a video as a link to it in a tweet.
- Upload and sharing of a photo.

(5) Comparison with similar media

Twitter is text-oriented like general blogging platforms such as WordPress [WordPress 2014] and Blogger [Blogger 2014]. Of course, tweets can also include links to other media as described above. On the other hand, the number of characters of tweets is less than that of general blog articles and tweets are more frequently posted. Incidentally, WordPress is not only a platform of blogging, but it also enables easy construction of applications upon LAMP (Linux Apache MySQL PHP) stacks, therefore it is widely used as CMS (Content Management System) for enterprises.

(6) API

Twitter offers REST (Representational State Transfer) and streaming as its Web services API.

1.2.2 Flickr

(1) Category and foundation

Flickr [Flickr 2014] [Flickr–Wikipedia 2014] is a photo sharing service launched by Ludicorp, a company founded by Stewart Butterfield and Caterina Fake in 2004 (see Fig. 1.2). Flickr focused on a chat service with

Figure 1.2 Flickr.

Color image of this figure appears in the color plate section at the end of the book.

real-time photo exchange in its early stages. However, the photo sharing service became more popular and the chat service, which was originally the main purpose, disappeared, partly because it had some problems.

(2) Numbers

- Registered users: 87 M
- The number of photos: 6 B

(3) Data structures

(Related to user)
- Account
- Profile

(Related to contents)
- Photo
- Set collection of photos
- Favorite photo
- Note
- Tag
- Exif (Exchangeable image file format)

(Related to relationships)
- Group
- Contact
- Bookmark of an album (a photo)

(4) Main interactions

- Creation and deletion of an account.
- Creation and change of a profile.
- Upload of a photo.
- Packing photos into a set collection.
- Appending notes to a photo.
- Arranging a photo on a map.
- Addition of a photo to a group.
- Making relationships between friends or families from contact.
- Search by explanation and tag.

(5) Comparisons with similar media

Although Picasa [Picasa 2014] and Photobucket [Photobucket 2014] are also popular like Flickr in the category of photo sharing services, here we will take up Pinterest [Pinterest 2014] and Instagram [Instagram 2014] as new players which have unique features. Pinterest provides lightweight services on the user side compared with Flickr. That is, in Pinterest, the users can not only upload original photos like Flickr, but can also stick their favorite photos on their own bulletin boards by pins, which they have searched and found on Pinterest as well as on the Web. On the other hand, Instagram offers the users many filters by which they can edit photos easily. In June, 2012, an announcement was made that Facebook acquired Instagram.

(6) API

Flickr offers REST, XML-RPC (XML-Remote Procedure Call), and SOAP (originally, Simple Object Access Protocol) as Web service API.

1.2.3 YouTube

(1) Category and foundation

YouTube [YouTube 2014] [YouTube–Wikipedia 2014] is a video sharing service founded by Chad Hurley, Steve Chen, Jawed Karim, and others in 2005 (see Fig. 1.3). When they found difficulties in sharing videos which had recorded a dinner party, they came up with the idea of YouTube as a simple solution.

Figure 1.3 YouTube.

Color image of this figure appears in the color plate section at the end of the book.

(2) Numbers

- 100 hours of movies are uploaded every minute.
- More than 6 billion hours of movies are played per month.
- More than 1 billion users access per month.

(3) Data structures

(Related to users)
- Account

(Related to contents)
- Video
- Favorite

(Related to relationships)
- Channel

(4) Main interactions

- Creation and deletion of an account
- Creation and change of a profile
- Uploading a video
- Editing a video
- Attachment of a note to a video
- Playing a video
- Searching and browsing a video
- Star-rating of a video
- Addition of a comment to a video
- Registration of a channel in a list
- Addition of a video to favorite
- Sharing of a video through e-mail and SNS

(5) Comparison with similar media

As characteristic rivals, Japan-based Niconico (meaning smile in Japanese) [Niconico 2014] and the US-based USTREAM [USTREAM 2014] are picked up in this category. Although the Niconico Douga, one of the services provided by Niconico, is similar to YouTube, Niconico Douga allows the user to add comments to movies which can be superimposed on the movies and seen by other users later, unlike YouTube. Such comments in Niconico Douga have attracted a lot of users as well as the original contents. Niconico Live is another service provided by Niconico and is similar to the live video service of USTREAM. USTREAM was originally devised as a way by which US soldiers serving in the war with Iraq could communicate with their families. The function for posting tweets simultaneously with video viewing made USTREAM popular. Both USTREAM and Niconico Live can be viewed as a new generation of broadcast services which are more targeted than the conventional mainstream services.

(6) API

YouTube provides the users with a library which enables the users to invoke its Web services from programming environments, such as Java and PHP.

1.2.4 Facebook

(1) Category and foundation

Facebook [Facebook 2014] [Facebook–Wikipedia 2014] is an integrated social networking service founded by Mark Zuckerberg and others in 2004, where the users participate in social networking under their real names (see Fig. 1.4). Facebook began from a site which was intended to promote exchange among students and has since grown to be a site which may affect fates of countries. Facebook has successfully promoted development of applications for Facebook by opening wide its development platform to application developers or giving them subsidies. Furthermore, Facebook has invented a mechanism called social advertisements. By Facebook's social advertisements, for example, the recommendation "your friend F

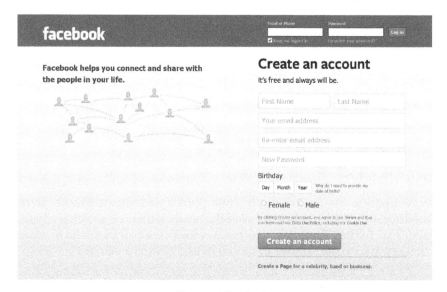

Figure 1.4 Facebook.

Color image of this figure appears in the color plate section at the end of the book.

purchased the product P" will appear on the page of the user who is a friend of F. Facebook's social advertisements are distinguished from anonymous recommendations based on historical mining of customers behavior such as that of Amazon.

(2) Numbers

- Active users: 1 B

(3) Data structures

(Related to user)
- Account
- Profile

(Related to contents)
- Photo
- Video

(Related to relationships)
- Friend list
- News feed
- Group

(4) Main interactions

- Creation and deletion of an account
- Creation and update of a profile
- Friend search
- Division of friends into lists
- Connection search
- Contribution (recent status, photo, video, question)
- Display of time line
- Sending and receiving of a message

(5) Comparison with similar media

In addition to the facilities to include photos and videos like Flickr or YouTube, Facebook has also adopted the timeline function which is a basic facility of Twitter. Facebook incorporates the best of social media in other categories, so to say, a more advanced hybrid SNS as a whole.

(6) API

FQL (Facebook Query Language) is provided as API for accessing open graphs (that is, social graphs).

1.2.5 Wikipedia

(1) Category and foundation

Wikipedia [Wikipedia 2014] is an online encyclopedia service which is a result of collaborative work, founded by Jimmy Wales and Larry Sanger in 2001 (see Fig. 1.5). The history of Wikipedia began from Nupedia [Nupedia 2014], a project prior to it in 2000. Nupedia aimed at a similar online encyclopedia based on copyright-free contents. Unlike Wikipedia, however, Nupedia had adopted the traditional editorial processes for publishing articles based on the contributions and peer reviews by specialists. Originally, Wikipedia was constructed by a Wiki software for the purpose of increasing articles as well as contributors for Nupedia in 2001. In the early stages of Wikipedia, it earned its repulation through electric word-of-mouth and attracted a lot of attention through being mentioned in Slashdot

Figure 1.5 Wikipedia.

[Slashdot 2014], a social news site. Wikipedia has rapidly expanded its visitor attraction with the aid of search engines such as Google.

(2) Numbers

- Number of articles: 4 M (English-language edition)
- Number of users: 20 + M (English-language edition)

(3) Data structures

(Related to users)
- Account

(Related to contents)
- Page

(Related to relationship)
- Link

(4) Main interactions

(Administrator or editor)
- Creation, update, and deletion of an article
- Creation, update, and deletion of link to an article
- Change management (a revision history, difference)
- Search
- User management

(General user)
- Browse Pages in the site
- Search Pages in the site

(5) Comparison with similar media

From a viewpoint of platforms for collaboration, Wikipedia probably should be compared with other wiki media or cloud services (e.g., ZOHO [ZOHO 2014]). However, from another viewpoint of "search of knowledge" as the ultimate purpose of Wikipedia, players for social search services will be rivals of Wikipedia. You should note that differences between major search engines (e.g., Google [Google 2014] and Bing [Bing 2014]) and Wikipedia is being narrowed. Conventionally, such conventional search engines mechanically rank the search results and display them to the users. However, by allowing the users to intervene between search processes in certain forms, the current search engines are going to improve the quality of search results. Some search engines include relevant pages linked by friends in social media in search results. In order to get answers to a query, other search engines discover people likely to answer the query from friends in social media or specialists on the Web, based on their profiles, uploaded photos, and blog articles.

(6) API

In Wikipedia, REST API of MediaWiki [MediaWiki 2014] can be used for accessing the Web services.

1.2.6 Generic Web

(1) Category and foundation

When Tim Berners-Lee joined CERN as a fellow, he came up with the prototype of the Web as a mechanism for global information sharing and created the first Web page in 1990. The next year, the outline of the WWW project was released and its service was started. Since the Web, in a certain sense, is the entire world in which we are interested, it contains all the categories of social media.

(2) Numbers

- The size of the indexable Web: 11.5 + B [Gulli et al. 2005]

(3) Data Structures

(Related to users)
NA
(Related to contents)
- Page
(Related to relationships)
- Link

(4) Main operations

(Administrator)
- Creation, update, and deletion of a page
- Creation, update, and deletion of a link
(General user)
- Page browse in a site
- Page search in a site
- Form input

(5) Comparisons with similar media

Since the Web is a universal set containing all the categories, we cannot compare it with other categories. Generally, the Web can be classified into the surface Web and the deep Web. While the sites of the surface Web allow the user to basically follow links and scan pages, those of the deep Web with back-end databases, create pages dynamically and display them to the user, based on the result of the database query which the user issues through the search form. Moreover, the sites of the deep Web are increasing rapidly [He et al. 2007]. The categories in the deep Web include on-line shopping services represented by Amazon, and various kinds of social media described in this book.

(6) API

Web services API provided by search engines such as Yahoo! can facilitate search of Web pages. Unless we use such API, we need to carry tedious Web crawling by ourselves.

1.2.7 Other social media

The categories of social media which have not yet been discussed will be enumerated below.

- Sharing service: In addition to photos and videos described previously, audios (e.g., Rhapsody [Rhapsody 2014], iTunes [iTunes 2014]) and bookmarks (e.g., Delicious [Delicious 2014], Japan-based Hatena bookmark [Hatena 2014]) are shared by users.
- Video communication: Users can communicate with each other through live videos. Skype [Skype 2014] and Tango [Tango 2014] are included in this category.
- Social news: The users can post original news or repost existing news by adding comments to them. Representative media of this category include Digg [Digg 2014] and Reddit [Reddit 2014] in addition to Slashdot.
- Social gaming: A group of users can play online games. The services in this category include FarmVille [FarmVille 2014] and Mafia Wars [Mafia Wars 2014].
- Crowd sourcing: The services in this category allow personal or enterprise users to outsource the whole or parts of a job to crowds in online communities. Amazon Mechanical Turk [Amazon Mechanical Turk 2014] for requesting labor-oriented work and InnoCentive [InnoCentive 2014] for requesting R&D-oriented work are included by the services in this category.

References

[Amazon Mechanical Turk 2014] Amazon Mechanical Turk: Artificial Intelligence https://www.mturk.com/mturk/welcome accessed 2014

[Bing 2014] Bing http://www.bing.com accessed 2014

[Blogger 2014] Blogger https://www.blogger.com accessed 2014

[Delicious 2014] Delicious http://delicious.com accessed 2014

[Digg 2014] Digg http://digg.com accessed 2014

[Facebook 2014] Facebook https://www.facebook.com/accessed 2014

[Facebook–Wikipedia 2014] Facebook–Wikipedia http://en.wikipedia.org/wiki/Facebook accessed 2014

[FarmVille 2014] FarmVille http://company.zynga.com/games/farmville accessed 2014

[Flickr 2014] Flickr https://www.flickr.com accessed 2014

[Flickr–Wikipedia 2014] Flickr–Wikipedia http://en.wikipedia.org/wiki/Flickr accessed 2014

[Google 2014] Google https://www.google.com accessed 2014

[Gulli et al. 2005] A. Gulli and A. Signorini: The indexable web is more than 11.5 billion pages. In Special interest tracks and posters of the 14th international conference on World Wide Web (WWW '05). ACM 902–903 (2005).

[Hatena 2014] Hatena http://www.hatena.ne.jp/accessed 2014

[He et al. 2007] Bin He, Mitesh Patel, Zhen Zhang and Kevin Chen-Chuan Chang: Accessing the deep web, Communications of the ACM 50(5): 94–101 (2007).

[InnoCentive 2014] InnoCentive http://www.innocentive.com accessed 2014

[Instagram 2014] Instagram http://instagram.com/accessed 2014

[iTunes 2014] iTunes https://www.apple.com/itunes/accessed 2014

[Mafia Wars 2014] Mafia Wars http://www.mafiawars.com/accessed 2014

[MediaWiki 2014] MediaWiki http://www.mediawiki.org/wiki/MediaWiki accessed 2014

[Niconico 2014] Niconico http://www.nicovideo.jp/?header accessed 2014

[Nupedia 2014] Nupedia http://en.wikipedia.org/wiki/Nupedia accessed 2014

[Picasa 2014] Picasa https://www.picasa.google.com accessed 2014

[Photobucket 2014] Photobucket http://photobucket.com/accessed 2014

[Pinterest 2014] Pinterest https://www.pinterest.com/accessed 2014

[Reddit 2014] Reddit http://www.reddit.com accessed 2014

[Rhapsody 2014] Rhapsody http://try.rhapsody.com/accessed 2014

[Skype 2014] Skype http://skype.com accessed 2014

[Slashdot 2014] Slashdot http://www.slashdot.org accessed 2014

[Tango 2014] Tango http://www.tango.me accessed 2014

[Twitter 2014] Twitter https://twitter.com accessed 2014

[Twitter-Wikipedia 2014] Twitter-Wikipedia http://en.wikipedia.org/wiki/Twitter accessed 2014

[USTREAM 2014] USTREAM http://www.ustrea.tv accessed 2014

[Wikipedia 2014] Wikipedia https://wikipedia.org accessed 2014

[WordPress 2014] WordPress https://wordpress.com accessed 2014

[YouTube 2014] YouTube http://www.youtube.com accessed 2014

[YouTube–Wikipedia 2014] YouTube–Wikipedia http://en.wikipedia.org/wiki/YouTube accessed 2014

[ZOHO 2014] ZOHO https://www.zoho.com/accessed 2014

Big Data and Social Data

At this moment, data deluge is continuously producing a large amount of data in various sectors of modern society. Such data are called big data. Big data contain data originating both in our physical real world and in social media. If both kinds of data are analyzed in a mutually related fashion, values which cannot be acquired only by independent analysis will be discovered and utilized in various applications ranging from business to science. In this chapter, modeling and analyzing interactions involving both the physical real world and social media as well as the technology enabling them will be explained. Data mining required for analysis will be explained in Part II.

2.1 Big Data

In the present age, large amounts of data are produced every moment in various fields, such as science, Internet, and physical systems. Such phenomena collectively called data deluge [Mcfedries 2011]. According to researches carried out by IDC [IDC 2008, IDC 2012], the size of data which are generated and reproduced all over the world every year is estimated to be 161 exa bytes (see Fig. 2.1). Here, kilo, mega, giga, tera, peta, exa, zetta are metric prefixes that increase by a factor of 10^3. Exa and Zetta are the 18th power of 10 and the 21st power of 10, respectively. It is predicted that the total amount of data produced in 2011 exceeded 10 or more times the storage capacity of the storage media available in that year.

Astronomy, environmental science, particle physics, life science, and medical science are among the fields of science which produce a large amount of data by observation and analysis of the target phenomena. Radio telescopes, artificial satellites, particle accelerators, DNA sequencers, and MRIs continuously provide scientists with a tremendous amount of data.

Nowadays, even ordinary people, not to mention experts, produce a large amount of data directly and intentionally through the Internet

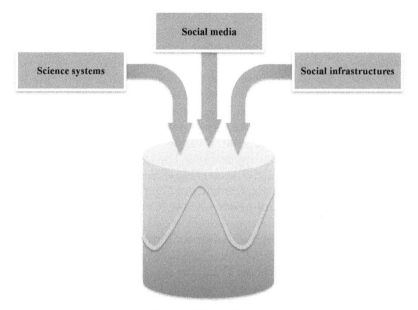

Figure 2.1 Data deluge.

services. Digital libraries, news, Web, Wiki, and social media. Twitter, Flickr, Facebook, and YouTube are representatives of the social media which have evolved rapidly in recent years. Moreover, some news sites such as Slashdot and some Wikis such as Wikipedia can be viewed as kinds of social media.

On the other hand, data originating in electric power apparatus, gas apparatus, digital cameras, surveillance video cameras, and sensors within buildings (e.g., passive infrared-, temperature-, illuminance-, humidity-, and carbon dioxide-sensors) and data originating in transportation systems (e.g., means of transportation and logistics) are among the data which people produce indirectly and unconsciously in physical systems. Until now, such data produced by physical systems was considered, so to speak, as data exhaust [Zikopoulos et al. 2011] of people. Nowadays, however, it is thought that it is possible to recycle such data exhaust and to generate business values out of them.

In the report of the above mentioned researches of IDC, data produced in science, the Internet, and physical systems are collectively called big data.

The features of big data can be summarized as follows:

- The quantity (Volume) of data is extraordinary, as the name denotes.
- The kinds (Variety) of data have expanded into unstructured texts, semi-structured data such as XML, and graphs (i.e., networks).
- As is often the case with Twitter and sensor data streams, the speed (Velocity) at which data are generated is very high.

Therefore, big data are often characterized as V^3 by taking the initial letters of these three terms Volume, Variety, and Velocity. Big data are expected to create not only knowledge in science but also values in various businesses.

By variety, the author of this book means that big data appear in a wide variety of applications. Big data inherently contain "vagueness" such as inconsistency and deficiency. Such vagueness must be resolved in order to obtain quality analysis results. Moreover, a recent survey done in Japan has made it clear that a lot of users have "vague" concerns as to the securities and mechanisms of big data applications. The resolution of such concerns are one of keys to successful diffusion of big data applications. In this sense, V^4 should be used for the characteristics of big data, instead of V^3.

Social media data are a kind of big data that satisfy these V^4 characteristics as follows: First, sizes of social media are very large, as described in chapter one. Second, tweets consist mainly of texts, Wiki media consist of XML (semi-structured data), and Facebook articles contain photos and movies in addition to texts. Third, the relationships between users of social media, such as Twitter and Facebook, constitute large-scale graphs (networks). Furthermore, the speed of production of tweets is very fast. Social data can also be used in combination with various kinds of big data though they inherently contain contradictions and deficits. As social data include information about individuals, sufficient privacy protection and security management are mandatory.

Techniques and tools used to discover interesting patterns (values) from a large amount of data include data mining such as association rule mining, clustering, and classification. On the other hand, techniques used to mainly predict occurrences of the future, using past data, include data analysis such as multivariate analysis [Kline 2011].

Of course, data mining and data analysis must more and more frequently treat such big data from now on. Therefore, even if data volume increases, data mining algorithms are required to be executable in practical processing time by systems realizing the algorithms. If the processing time of an algorithm increases proportionally as the data volume increases, then the algorithm is said to have linearity with respect to processing time. In other words, linearity means that it is possible that processing time can be maintained within practical limits by some means even if data volume increases. If an algorithm or its implementation can maintain such linearity by certain methods, then the algorithm or implementation is said to have scalability. How to attain scalability is one of the urgent issues for data mining and data analysis.

Approaches to scalability are roughly divided into the following: scale-up and scale-out. The former approach raises the processing capability (i.e., CPU) of the present computers among computing resources. On the other hand, the latter keeps the capability of each present computer as it is and instead multiplexes the computers. Internet giants, such as Amazon and Google, who provide large-scale services on the Internet, usually take scale-out approaches.

Next, with respect to the performance of processing large-scale data, there is another issue of high dimensionality in addition to scalability. Target data of data mining and data analysis can be viewed in many cases as objects consisting of a large number of attributes or vectors of a large number of dimensions. For example, depending on applications, the number of attributes and the dimension of vectors may be tremendously large such as feature vectors of documents, as described later. Issues which occur with the increase in the number of dimensions are collectively called a curse of dimensionality. For example, when sample data are to be collected at a fixed ratio for each dimension, there occurs a problem that the size of samples increases exponentially as data dimensionality increases. It is necessary for data mining and data analysis to appropriately treat even such cases.

Problems which data mining and data analysis must take into consideration are not confined only to the increase of data volume and that of data dimensionality. The complexity of the data structures to be handled also causes problems as application fields spread. Although conventionally data analysis and data mining have mainly targeted structured data such as business transactions, opportunities to handle graphs and semi-structured data are increasing along with the development of the Internet and Web. Moreover, sensor networks produce essentially time series data and GPS (Global Positioning System) devices can add location information to data. Unstructured multimedia data, such as photographs, movies, and audios, have also become the targets of data mining. Furthermore, in case the target data of data mining and data analysis are managed in a distributed fashion, problems such as communication costs, data integration and securities may occur in addition to the problems of complex data structures.

Please note that the term data deluge is just the name of a phenomena. In this book, the term big data will be used to mean more general concepts of large-scale data as well as analysis and utilization of them, but not the name of a phenomena. More precisely, this book will introduce an emerging discipline called social big data science and describe its concepts, techniques, and applications.

2.2 Interactions between the Physical Real World and Social Media

Based on the origins of where big data are produced, they can be roughly classified into physical real world data (i.e., heterogeneous data such as science data, event data, and transportation data) and social data (i.e., social media data such as Twitter articles and Flickr photos).

Most of the physical real world data are generated by customers who leave their behavioral logs in the information systems. For example, data about the customers check-in and check-out are inserted into the databases in the transportation management systems through their IC cards. Data about the customers use of facilities are also stored in the facility management databases. Further, the customers behaviors are recorded as sensor data and video data. In other words, real world physical data mostly contain only latent or implicit semantics because the customers are unconscious of their data being collected.

On the other hand, the customers consciously record their behaviors in the physical real world as social data on their own. For example, they post photos and videos, taken during events or trips, to sharing services and post various information (e.g., actions and sentiments) about the events or trips to microblogs. In a word, unlike physical real world data, social data contain explicit semantics because the customers voluntarily create the data.

Furthermore, there are bidirectional interactions between the physical real world data and social data through users (see Fig. 2.2). That is, if one direction of such interactions is focused on, it will be observed that events which produce physical real world data affect the users and make them describe the events in social data. Moreover, if attention is paid to the reverse direction of such interactions, it will turn out that the contents of social data affect other users actions (e.g., consumer behaviors), which, in turn, produce new physical real world data.

If such interactions can be analyzed in an integrated fashion, it is possible to apply the results to a wide range of application domains including business and science. That is, if interactions are analyzed paying

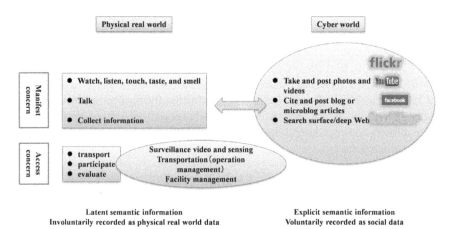

Figure 2.2 Physical real world data and social data.

Color image of this figure appears in the color plate section at the end of the book.

attention to the direction from physical real world data to social data, for example, the following can be accomplished.

- Measurement of effectiveness of marketing such as promotions of new products
- Discovery of reasons for sudden increase in product sales
- Awareness of need of measures against problems about products or services

Moreover, the following may be predicted if interactions are analyzed paying attention to the reverse direction of such interactions.

- Customer behaviors of the future
- Latent customer demands

All the above interactions are associated with applications which contain direct or indirect cause-effect relationships between physical real world data and social data. On the other hand, even if there exist no true correlations nor true causalities between both kinds of data, analysis of some interactions is useful.

For example, consider a situation where people go to a concert of a popular singer. After the concert, the people rush to the nearest train station resulting in the stations and trains getting crowded, as is often the case with the Japanese, among whom public transportation is more popular than automobiles. Through IC cards, the situation is recorded as traffic data, a kind of physical real world data in transportation. If the concert impresses the people, they will post a lot of articles to social media (see Fig. 2.3).

Those who are engaged in operations of transportation want to know reasons for the sudden increase (i.e., burst) in traffic data. However, it is not possible to know the reason only by analysis of traffic data. As previously described, physical real world data in general contain no explicit semantics. On the other hand, if social data posted near the stations after the concert can be analyzed, a sudden increase (i.e., another burst) in articles posted to the social media can be detected and then information about the concert can be extracted as main interests from the collection of articles. As a result, they will be able to conjecture that the people who attended the concert, caused the burst in traffic data. Like this case, some explicit semantics which are latent in physical real world data can be discovered from related social data.

Of course, there exist no cause-effect relationships (i.e., true correlations) between the two kinds of big data in the above case. In a word, participation in a concert gives rise to simultaneous increases in heterogeneous data (i.e., traffic data and social data) as a common cause. Thus, there exist *spurious* correlations between the two kinds of data. There, even if a true cause (e.g., concert participation) is unavailable, if such spurious correlations are positively utilized, one kind of data corresponding to another can

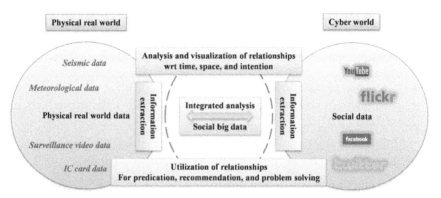

Figure 2.3 Integrated analysis of physical real world data and social data.

Color image of this figure appears in the color plate section at the end of the book.

be discovered. Such discovery enables the operation managers to take appropriate measures (e.g., distribution of clients to different stations) against future similar events (e.g., future concerts).

Of course, interactions may exist only within physical real world data or only within social data. The former contains data which are the targets of conventional data analysis such as causal relationships of natural phenomena. The latter contains upsurges in topics which are frequently argued only within social media.

Indeed, there are also some values in the analysis of such cases. However, cases where both physical real world data and social data are involved, are more interesting from a viewpoint of usefulness to businesses using social data. If physical real world data and social data can be analyzed by relating one to the other, and by paying attention to the interactions between the two, it may be possible to understand what cannot be understood, by analysis of only either of them. For example, even if only sales data are deeply analyzed, reasons for a sudden increase in sales, that is, what has made customers purchase more products suddenly, cannot be known. By analysis of only social data, it is impossible to know how much they contributed to sales, if any. However, if both sales data and social data can be analyzed by relating them to each other, it is possible to discover why items have begun to sell suddenly, and to predict how much they will sell in the future, based on the results. In a word, such integrated analysis is expected to produce bigger values.

Please note that the term social big data is frequently used throughout this book. Its intention is to make an emphasis on heterogeneous data sources including both social data and physical real world data as main targets of analysis.

2.3 Integrated Framework

In this section, from a viewpoint of hypotheses, we discuss the necessity of an integrated framework for analyzing social big data, which is beyond conventional approaches based on single use of either data analysis or data mining. In order to quantitatively understand physical real world data by using social data as mediators, quantitative data analysis such as multivariate analysis is necessary. In multivariate analysis, first, hypotheses are made in advance and then they are quantitatively confirmed. In other words, hypotheses play a central role in multivariate analysis. Conventionally, most models for hypotheses provide methods for quantitative analysis.

The importance of hypotheses does not change even in the big data era. However, the number of variables in big data may become enormous. In such cases, it is rather difficult to grasp the whole picture of the analysis. In other words, the problem of the curse of dimensionality occurs at the conceptual layer, too. The problem must be solved by hypothesis modeling.

Of course, since social data are a kind of big data, the volume of social data and the number of themes within social data are huge. However social data are sometimes very few or qualitative depending on individual themes and contents. For example, such data correspond to articles about minor themes or emerging themes. In such cases, not quantitative analysis but qualitative analysis is needed. That is, although quantitative confirmation of hypotheses cannot be performed, it is important to build and use qualitative hypotheses for explanation of phenomena.

Analyzing the contents of social data mainly requires data mining. Hypotheses also have an important role in data mining. Each task of data mining makes a hypothesis itself while each task of multivariate analysis verifies a given hypothesis. Therefore, it is desirable if the user (i.e., analyst) can give useful hints for making interesting hypotheses in each task to data mining systems.

In case of classification, it is necessary to allow the user to partially guide construction of hypotheses by selecting interesting data attributes (i.e., variables) or by showing empirical rules which can be fed to ensemble learning for final results. In case of clustering, it is necessary to enable the user to partially guide hypothesis construction by specifying individual data which must belong to the same cluster or general constraints which must be satisfied by data as members of the same cluster. It is also desirable to enable the user to specify parameters for clustering algorithms, constraints on whole clusters, and the definition of similarity of data in order to obtain clustering results interesting to the user. In case of association rule mining, it is necessary to guess items interesting to the user and required minimum support and confidence from concrete rules illustrated by the user as empirical knowledge. The above hints specified by the user are, so to say, early-stage hypotheses because they are helpful in generating hypotheses in the later stages of data mining.

In this book, analyzing both physical real world data and social data by relating them to each other, is called social big data science or social big data for short. To the knowledge of the author, there is no modeling framework which allows the end user or analyst to describe hypotheses spanning across data mining, quantitative analysis, and qualitative analysis. In other words, conceptual hypothesis modeling is required which allows the user to describe hypotheses for social big data in an integrated manner at the conceptual layer and translate them for execution by existing techniques such as multivariate analysis and data mining at the logical layer if needed.

By the way, a database management system, which is often used to store target data for mining, consists of three layers: the conceptual, logical, and physical layers. Following the three layered architecture of the database management system, the reference architecture of the integrated system for social big data science is shown in the Fig. 2.4. At the conceptual layer the system allows the user (i.e., analyst) to describe integrated hypotheses relating to social big data. At the logical layer, the system converts the hypotheses defined at the conceptual layer in order for the user to actually confirm them by applying individual techniques such as data mining and multivariate analysis. At the physical layer, the system performs further analysis efficiently by using both software and hardware frameworks for parallel distributed processing.

Here we introduce a conceptual framework for modeling interactions between physical real world data and social data. The introduced framework is called MiPS (Modeling interactions between Physical real world and Social media). Although the MiPS model has not yet been actually implemented, it will be used as a formalism for describing specific examples of integrated hypotheses in this book.

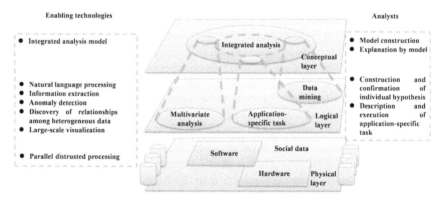

Figure 2.4 Reference architecture for social big data.

Color image of this figure appears in the color plate section at the end of the book.

2.4 Modeling and Analyzing Interactions

In this section the procedure of modeling and analyzing interactions between the physical real world and social media will be explained.

Generally, the procedure is performed step by step as follows (see Fig. 2.5):

- (Step one) Setup of problem
- (Step two) Modeling of interactions between physical real world and social media (hypothesis construction)
- (Step three) Collection of data
  i. Extraction of information from physical real world data
  ii. Extraction of information from social data
- (Step four) Analysis of influences of physical real world on social media (hypothesis confirmation 1)
- (Step five) Analysis of influences of social media on the physical real world (hypothesis confirmation 2)
- (Step six) The bidirectional analysis by integrating influences described in the steps four and five in order to complete the whole model (theory) as explanation of the interactions.

①	Setup of problem
②	Modeling of interactions between physical real world and social media (hypothesis construction)
③	Collection of data
	I. Extraction of information from physical real world data
	II. Extraction of information from social data
④	Analysis of influences of physical real world on social media (Hypothesis confirmation 1)
⑤	Analysis of influences of social media on the physical real world (Hypothesis confirmation 2)
⑥	The bidirectional analysis by integrating influences described in the steps *Four* and *Five* in order to complete the whole model (theory) as explanation of the interactions.

Figure 2.5 Analytic procedure.

There may be feedback, if needed, from each step to the precursor step. Some application domains require only either of the steps four and five in the procedures described above. Moreover, the order of these steps are determined depending on application domains. Each step in the procedure will be described in more detail below.

(1) Problem setup

In step one the user sets up problems to be solved. Such problems can often be formulated in the form of questions. In other words, at this stage, the user describes a phenomenon of interest in a certain area at a certain time in order to explain it to others. The basic types of questions vary, depending on analytical purposes as follows:

- Discover causes (Why did it happen?)
- Predict effects (What will happen?)
- Discover relationships (How are they related to each other?)
- Classify data into known categories (To which category does it belong?)
- Group data that are similar to each other (How similar are they to each other?)
- Find exceptions (How seldom is it?)

In the sense that these questions help the user to roughly determine types of analytical tasks the user should perform hereafter, it is very important to focus on the purpose of the question. Further, in order to solve the problem, the user clearly defines the requirements as to what data to use, what kind of analytical technique to apply, and what criteria for the hypothesis to adopt.

(2) Hypothesis construction

In step two, the user constructs a hypothesis as a tentative solution to the problem. To this end, this book proposes a framework that focuses on the relationships between social data and physical real world data and conceptually models them in an object-oriented manner. Please note that relationships between heterogeneous physical real world data are modeled if necessary. Indeed, there are some approaches which support graphical analysis of related variables in multivariate analysis. However, they are, so to say, value-oriented, that is, fine grained. In contrast, hypothetical modeling proposed in this book are based on the relationships between objects in a more coarse grained fashion. Physical events, such as product campaigns and earthquakes, and the contents of tweets, such as product evaluation and earthquake feedback, are considered as first class objects, which are called big objects. In the proposed model, inherently related variables are grouped into one big object and represented as attributes of the big object. For example, in case of an earthquake, the epicenter and magnitude of the earthquake as objective values or the subjective intensity at a place where the earthquake was felt as well as the date and time when the earthquake happened or was felt are considered as attributes of the big object earthquake while in case of a marketing campaign, the name and reputation of the product and the type and cost of the campaign are considered as attributes of the campaign big object. An influence relationship between two big objects (not variables) is collectively described as one or more causal relationships between the variables (attributes) of the objects. Once these models are built, in the rest of the above procedure the user is able to analyze the subject based on the big picture drawn as big objects and the relationships between them.

Structural Equation Modeling (SEM) is among the multivariate analysis techniques that describe causal relationships between variables by introducing latent factors. It is possible to correspond the latent factors that are identified by SEM to candidate big objects in the proposed framework. However, the analytical model proposed in this book is independent of the existing analytical techniques. In other words, the proposal is a framework for conceptual analysis and can coexist with logical and operational analytical techniques such as multivariate analysis and data mining. For short, conceptual analysis models constructed in the framework will be converted into logical analysis models for execution by the actual analysis methods.

In the classification task of data mining, an influence relationship is described as a directed relationship from a big object with classification attributes to another big object with a categorical attribute. Such two big objects may be the same in a special case. In the clustering task, an influence relationship is described as a self-loop effect from one big object from the same object. Similarly, in association rule mining, an influence relationship is described as a self-loop effect from and to one big object.

The model proposed in this book will be used as a meta-analysis model as follows. Prior to detailed analysis of the interactions that are done in the subsequent steps four and five, in this stage the analyst (i.e., the expert user) instantiates this meta-analysis model and constructs specific hypotheses by combining the instances by using influence relationships between big objects in the field of applications. It goes without saying that the hypotheses are constructed using preceding theories and prior observations in addition to the required specifications and problem setting, too.

(3) Data collection

In step three, social big data required for analysis and confirmation of hypotheses constructed in the previous step are collected. Social data are collected either by searching or streaming through the API provided by the relevant sites and are stored in dedicated databases or repositories. As physical real world data are often collected in advance and stored in separate databases, necessary data are selected from the databases. After the data undergo appropriate data cleansing and optional data conversion, the data are imported into dedicated databases for analysis.

   i. Information extraction is performed on physical real world data. For example, remarkable events as interests of the users (i.e., analysts) are discovered from the data by using techniques such as outlier detection and burst detection.

  ii. Similarly, information extraction is performed on social data. For example, interests of the users (i.e., customers) are discovered from the data by applying text mining techniques to natural language contents and by applying density-based clustering to photos for detecting shooting directions.

(4) Hypothesis confirmation

In steps four and five, specific analysis methods, such as multivariate analysis and data mining, are applied to the collected data in order to discover causalities and correlations between them. Thus the primary hypotheses are confirmed. The analysts may modify the hypotheses (i.e., influence relationships between big objects) according to the results if necessary. It goes without saying that these two steps are performed not in a separate manner but in an integrated manner. Furthermore, hypotheses involving heterogeneous physical real world data are confirmed, if any.

In step six the hypotheses constructed in the previous steps are completed in order to be usable for final description of interactions. In other words, the completed hypotheses are upgraded to certain theories in the application domains at this time. The description of the hypotheses also requires large-scale visualization technology appropriate for big data applications. Large-scale visualization can also be used to obtain hints for building hypotheses themselves.

Hypotheses in the era of big data, in general, will be discussed later in more detail.

2.5 Meta-analysis Model—Conceptual Layer

The meta-analysis model, which is required throughout the whole procedure of analysis will be described in detail here. In an integrated framework for analysis of social big data, the meta-analysis model, which corresponds to classes for specific applications is instantiated and the instantiated model is used as an application-specific hypothetical model at the conceptual layer closest to the users. Although social media are not limited to Twitter, of course, Twitter will be used mainly as working examples throughout this book.

2.5.1 Object-oriented Model for Integrated Analysis

In this book, an integrated framework for describing and analyzing big data applications will be introduced. Unlike multivariate analysis, the purpose of the integrated model at the core of the framework is not confirmation of microscopic hypotheses, but construction and analysis of macroscopic hypotheses as well as high level description and explanation of social big data applications. The instantiated model is hereafter called the model.

One of the basic components of the model is a big object, which describes associated big data sources and tasks (see Fig. 2.6). Such tasks include construction of individual hypotheses related to big data sources (e.g., data mining), confirmation of the individual hypotheses (e.g., multivariate analysis), information extraction from natural language data, data monitoring or sensing, and other application-specific logics (programs). As another component of the model, influence relationships are described between big data objects. They represent causalities, correlations, and spurious (pseudo) correlations. Tasks can also be attached to influence relationships. Such tasks perform matching heterogeneous big data sources and detecting various relationships among them.

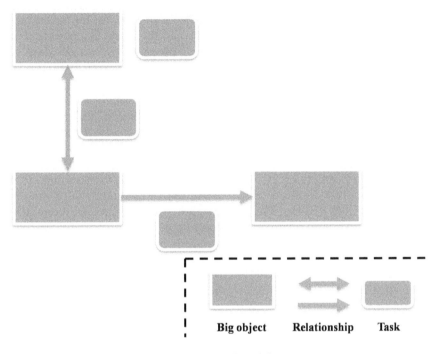

Figure 2.6 MiPS model.

The features of the model can be summarized as follows:

- Social big data applications are described in a high level fashion by using big objects and influence relationships between them.
- Big objects describe big data sources and tasks.
- Big data sources specify a set of inherently related big data.
- Tasks specify operations on big data sources in a high level fashion, which are refined for execution by specific analytical tools or data mining libraries.
- Influence relationships describe spurious correlations and qualitative causalities as well as correlations and quantitative causalities in a high level fashion.

- Tasks for discovering influence relationships are attached to the relationships. Such cases involve at least two big data sources. The tasks are refined for execution as well.
- The completed model explains the whole big data application and contributes to decrease in vague concerns about big data utilization among the users.

As introduced above, big objects, attributes, and relationships constitute elements for describing hypotheses. In step two, both social data and physical real world data are recognized as big objects. All the inherently related variables are defined as attributes of the same big objects. For example, an influence relationship from physical real world data to social data is expressed as one or more equations involving attributes of corresponding big objects. Such equations are usually expressed as linear functions that represent mappings between the attributes of the big objects. If there are relationships between the internal variables (i.e., attributes of the same big object), such relationships may be expressed as equations involving the attributes as well. If there is a prerequisite for influence relationships, such a prerequisite is represented by logical expressions as to variables. Equations and optional logical expressions constitute relationships. In a word, the analyst describes concrete influences as relationships among attributes of big objects. Please note that relationships are generally described as domain-dependent computational logics.

The mapping between the meta-analysis model introduced here and SEM (Structural Equation Modeling) is intuitively described in a case where the analyst wants to use SEM as a specific technique for multivariate analysis. Consider the following example, that is, multi-indicator model.

$X_1 = \lambda_{12}F_1 + e_1$: measurement equation
$X_2 = \lambda_{21}F_1 + e_2$
$X_3 = \lambda_{32}F_2 + e_3$
$X_4 = \lambda_{42}F_2 + e_4$
$F_2 = \lambda_{12}F_1 + d_2$: structural equation

Let big objects and their attributes correspond to latent factors in SEM model, such as F_1 and F_2, and to observable variables associated with the latent factors, such as X_1 and X_2, respectively. The big objects can have "special attributes" which represent the values of their own. In that case, the values of normal attributes (i.e., observed variables) are assumed to be calculated from the values of such special attributes (i.e., latent variables). They are collectively represented as a set of measurement equations. Influence relationships between the objects are represented by a set of structural equations between the special attributes of the objects, which correspond to latent factors.

Now let us consider a simpler model, that is, multiple regression analysis (including linear regression analysis) than SEM analysis in general. Let independent and dependent variables in this case correspond to attributes of big objects like SEM. Let's consider the following model.

$$X_3 = \gamma_{31}X_1 + \gamma_{32}X_2 + e_3$$

Here γ_{31} and γ_{32} denote path coefficients and e_3 denotes an error.

Some notes will be made about variables. The attributes corresponding to the effect variables (e.g., X_3) are represented as expressions of attributes corresponding to the cause variables (e.g., X_1 and X_2). In case of multiple regression analysis, it is not necessary to prepare special attributes introduced above for big objects in the case of SEM. If it is recognized that two variables belong to different entities in the physical real world even if there exist no other variables, they are to be represented as attributes of separate big objects.

In the classification task of data mining, the influence relationship is described from one big object with classification attributes to another big object with a categorical attribute. Of course, these objects can be the same in a special case. It is desirable that the user is able to illustrate empirical classification rules and specific attributes of interest prior to the task in order for the system to take them into consideration.

In case of clustering, the relationship is described as a self-loop from the target big object to the big object itself. Since the result of clustering is the sum of the partitioned subsets, the relationship is represented as, for example, "+". In this case, it is desirable that the user (i.e., analyst) can illustrate the combination of individual objects that must belong to the same cluster and the combination of individual objects that must belong to separate clusters by the enumeration of specific objects or the constraints between the objects.

In mining of association rules, the relationship is also described as a self-loop from one big object to the same big object in a similar way. In this case, since association rule mining is equivalent to discovering elements of the power set of a set of items S, the relationship between the big object is denoted by, for example, "2^S". In this case, it is desirable if the user can illustrate empirical association rules in addition to items of interest as examples as in the case of other tasks. Then the system will be able to guess minimum support and confidence of interest from the illustrated rules.

Relations between our integrated analysis model and data analysis such as SEM will be described. Some parts of the integrated model (i.e., big objects and influence relationships together with attached tasks) can be systematically translated into what data analysis tools such as SEM can analyze at the logical layer. However, big objects can also contain what should be analyzed by data mining tools or application-dependent logics as well as qualitative analysis methods. In other words, the integrated analysis model cannot necessarily be constructed and validated all at once. As many integrated hypotheses as possible should be constructed in a top-down manner and then each of them should be translated and validated by appropriate tools at the logical layer. Thus, the integrated analysis model is validated and completed as a whole.

2.5.2 Primitive Cases

As a meta-analysis model for describing and validating hypotheses, two or more primitive cases can be considered as shown in Fig. 2.7.

P (physical real world data) and S (social data) are big objects. Since P and S are classes, they will be denoted by capital letters here. Moreover, "->" expresses influence relationships.

- P->S: If an event with some impact occurs in the physical real world, it is described in social media.
- S->P: What is described in social media effects human behavior in the physical real world.

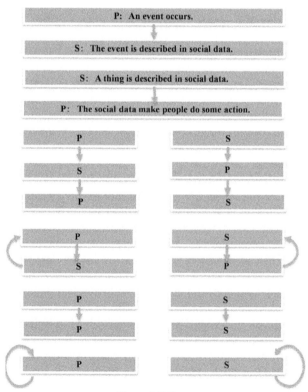

Figure 2.7 Cases.

- P->S->P or S->P->S: There are cases either with or without a loop.
- P->P: There are cases either with or without a loop.
- S->S There are cases either with or without a loop.

Hypotheses in reality are created by instantiating the classes P and S and combining the instances. The hypotheses which include, in particular, both instances of the classes S and P are interesting from a viewpoint of business applications in this book although the framework can describe sufficiently general cases. Influence relationships herein represent not mere causality but its extended concept. In other words, while causal relationships are suitable for the microscopic analysis of the relationships between variables of data, influence relationships are suitable for macroscopic analysis of the relationships between big objects.

2.6 Generation and Confirmation of Hypotheses—Logical Layer

The integrated hypotheses described at the conceptual layer of the framework are translated into those at the logical layer where hypothesis generation and confirmation are performed by using tools for data analysis and data mining. This subsection will explain multivariate analysis and data mining briefly.

2.6.1 Multivariate Analysis

Among the hypotheses described at the conceptual layer, portions corresponding to microscopic hypotheses in multivariate analysis are translated into those that can be analyzed by specific data analysis tools. Prediction as one of the main functions of multivariate analysis will be explained separately later.

2.6.2 Data Mining

Among the hypotheses described at the conceptual layer, portions which should be mapped into hypotheses in each task of data mining are translated into those for concrete mining tools to generate and confirm them. Some of the basic algorithms of data mining will be explained in detail separately later.

Furthermore, the portions which the users (i.e., data analysts) specify for each task in data mining in order to show their interests as examples are utilized by data mining so as to generate interesting hypotheses.

More specifically, the attributes specified for classification tasks by the users can be used to create classifiers which contain them. Moreover, the specified classification rules can be used in ensemble learning in data mining. As a result, hypotheses reflecting the user's (i.e., analyst's) interests are created. If any specific combination of instance objects or specific constraints about objects within the same cluster are illustrated by the user, such specifications are taken into account in the execution of clustering tasks and as a result the hypotheses reflecting the user's interests are created. If items of interest are shown by the user as examples for association rule mining, then generation of association rules containing the items is done on a priority basis. If specific association rules are specified as a piece of empirical knowledge, confidence and support of the rules can be used to estimate the minimum support and confidence which the user expects.

In a word, it is expected that the user's illustration of such examples enables each task not only to induce the hypotheses interesting to the user but also to narrow the search space of hypotheses and to reduce the processing time as a result.

2.6.3 Discovery and Identification of Influences

First in step four, it is necessary to detect any existence of influences from the physical real world data to the social data. The existence of influences can be discovered by observing the dynamic states (dynamics) of social media data. Taking Twitter as an example, it is possible to detect the existence by paying attention to the following dynamic states.

- The burst in the time series, or line time of tweets, i.e., rapid change in the number of tweets per unit time on the line time
- Rapid change of network structures, such as follow-follower relationships and retweet relationships

Social media other than Twitter will be described here, too. Let's consider photo sharing services such as Flickr. If it is observed in a certain area that the number of photos per unit grid (i.e., the photo density) are more than a specified threshold, popular landmarks can be discovered. Moreover, if temporal change of the density is taken into consideration, so-called emergent hot spots that have recently begun to attract attention can be discovered [Shirai et al. 2013].

If any existence of influences is checked in this way, then it is necessary as a next step, to identify the physical real world data which affected social data. If information extraction techniques are applied to the contents of a bag of Tweets on the line time, the heterogeneous information sources corresponding to the physical real-world data are automatically identified. Such heterogeneous information sources contain open access media such as Wikipedia and open access journals, limited access ones such as enterprise data, and personal stuff such as hands-on experiences. They substantially correspond to topics. Basically, frequent topics are focused on therein.

It is also possible to use social tags. The users of social media add social tags to social data. Thus, user-defined tags including hash tags in the case of Twitter and user-defined tags other than EXIF in the case of Flickr, correspond to such social tags respectively. EXIF data are automatically added to the photos as the shooting conditions. Since social tags denote topics explicitly in many cases, heterogeneous information sources corresponding to such topics can be quickly discovered by analyzing them. However, in some cases different topics have the same tag and in other cases the same tag has a time-varying meaning. Such problems should be solved during the analysis of social tags.

For example, both #jishin and #earthquake were actually used to mean "3.11 Earthquake in Japan" just after the earthquake. In the course of time, only #jishin was dominantly used.

2.6.4 Quantitative Measurement of Influences

After the discovery and identification of influences, the influences from social data to physical real world data (step four) and those from the latter to the former (step five) must be quantitatively measured. To do so, from a set of social data, such as tweets, the articles that are important and related to the topic must be discovered.

First, it is conceivable to use a sentiment polarity dictionary [Taboada et al. 2011] suitable for business topics that has been constructed in advance. The sentiment polarity dictionary is obtained by adding a certain value ranging from positive to negative as a polarity value to each word entry. The importance of an article is determined based on the polarity of the words contained by the article content, depending on application domains.

For example, to determine whether a new product campaign is successful or not, it is sufficient to analyze a set of articles in the entire spectrum of sentiment polarity. Articles that have values of negative sentiment polarity are mainly analyzed so as to analyze complaints and improvement of a product. In this case, however, a sufficient number of articles cannot always be collected for quantitative analysis. In such a case, it is necessary to analyze individual articles as objectively as possible by using the qualitative analysis method.

Basically in this way, the influences between physical real world data and social data are quantitatively analyzed. As already mentioned, of course, qualitative analysis techniques are also used if needed.

In step four, it is necessary to evaluate the articles (e.g., tweets) of social media with respect to the degrees of relevance to a particular topic, of accuracy of prediction about the topic, and of influence to other users.

For example, it is possible to use the following measures for such evaluations, respectively:

- The degree of the relevance and specialty of the articles about the topic
- The accuracy of predictions about the topic by the past articles of the contributor
- The size of the network consisting of follower-followee relationships of the contributor of the article

Text mining, graph mining, and multivariate analysis are performed to evaluate the article by using these measures. If articles similar to the past influential articles are detected by monitoring newly generated articles, it is expected that various predictions on businesses can be performed by analyzing the articles.

As with steps four and five among the analytic procedures, the scalability with respect to the size of big data is required for the performance of data mining algorithms. As one of such approaches, it is possible to extend data mining algorithms based on conventional single processors by using a platform for distributed parallel computing such as MapReduce which can work on a platform for distributed processing such as Hadoop. Hadoop and MapReduce will be explained below very briefly.

2.7 Interests Revisited—Interaction Mining

So far, only data analyst's interests have been discussed. There are other interests in addition to them. Needless to say, interests of customers are very important, which will be explained in this subsection.

Traditionally, data mining handles transactions which are recorded in databases if the customers actually purchase products or services. Analyzing transactional data leads to discovery of frequently purchased products or services, especially repeat customers. But transaction mining cannot obtain information about customers who are likely to be interested in products or services, but have not purchased any products or services yet. In other words, it is impossible to discover prospective customers who are likely to be new customers in the future.

In the physical real world, however, customers look at or touch interesting items displayed in the racks. They trial-listen to interesting videos or audios if they can. They may even smell or taste interesting items if possible and even if interesting items are unavailable for any reasons, customers talk about them or collect information about them.

These behaviors can be considered, as parts of interactions between customers and systems (i.e., information systems). Such interactions indicate the interests of latent customers, who either purchase interesting items or do not in the end, for some reasons.

Parts of such interactions are recorded in databases or repositories as video and sensor data through video cameras and sensors set inside stores, respectively. Through IC cards, customers leave check-in/check-out logs in facilities and means of transportation that they use in order to access products and services. These constitute big data in the physical real world. Such data include interactions or interests being accumulated, which customers are unaware of.

On the other hand, in the cyber world, users post photos or videos of interesting items (i.e., products or services) to social media such as Flickr and YouTube, which they have recorded. Some users mention interesting items in their own articles of blogs or microblogs such as Facebook and Twitter. Other users collect information about interesting items by searching the Web such as general pages, blogs or microblogs, Q&A sites, comparison shopping sites. These interactions are recorded as logs in the databases of the system. Such interactions in the cyber world accompany users' awareness, unlike those in the physical real world.

Analyzing interactions in the physical real world leads to understanding which items customers are interested in. By such analysis, however, which aspects of the items the customers are interested in, why they bought the items, or why they didn't, remain unknown. On the other hand, in social data, some users explicitly describe which parts of the items interested them, what caused them to buy or not buy the items. Therefore, if interests of the users are extracted from heterogeneous data sources and the reasons for purchasing items or the reasons for not purchasing the items are uncovered, it will be possible to obtain latent customers. In general, if the users' interests are extracted from heterogeneous big data sources and any matches between them (e.g., similarity or identity of interests) or relationships between them (e.g., causality from one interest to another or correlation between interests) are discovered, it is expected that more valuable information will be produced. In this book, traditional mining of transactional data and new mining of interactional data are distinctively called transaction mining and interaction mining, respectively (see Fig. 2.8).

A few remarks will be made about the two types of interests described above. While the analysts' interests are provided by the analysts themselves, the customers' interests should be discovered and analyzed by the systems. In special cases, both of them may coincide with each other. In other words, interaction mining can be allegorized to a journey where the users' interests are searched for as goals, guided by the analysts' interests as mile stones. In contrast, traditional transaction mining can be said to seek goals of model construction without any light of interests.

2.8 Distributed Parallel Computing Framework

The computing framework for analyzing social big data consists of multiple layers. The technologies and tools used at each layer contain the following:

The conceptual layer: This layer provides the big object model introduced in this chapter.

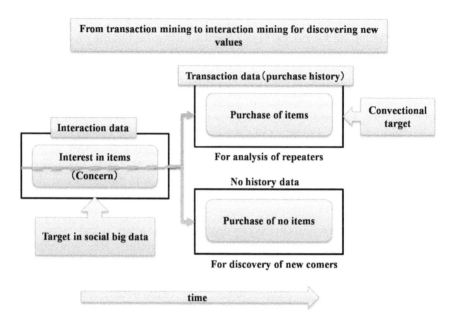

Figure 2.8 Interaction mining.

Color image of this figure appears in the color plate section at the end of the book.

The logical layer: This layer contains analytical tools such as multivariate analysis, data mining, machine learning, and natural language processing. Among these, in particular, data mining techniques, text mining techniques as one of natural language processing, and multivariate analysis are explained in separate chapters in Part II of this book. Machine learning is just briefly mentioned in relation to data mining.

The physical layer: This layer consists of software and hardware. For example, relational databases and NoSQL databases for data management, and Hadoop and MapReduce for distributed parallel computing are used as software. The outline of these will be briefly explained in the remainder of this section. As hardware, computer clusters are often used. As they exceed the scope of this book, they are not explained here.

2.8.1 NoSQL as a Database

It has been reported that 65% of queries processed by Amazon depend on primary keys (i.e., record identifiers) [Vogels 2007]. Therefore, as a mechanism for data access based on keys, key value stores [DeCandia et al. 2007] will be explained, which are currently data administration facilities used by Internet giants such as Google and Amazon.

The concrete key value stores include DynamoDB [DynamoDB 2014] of Amazon, BigTable [Chang et al. 2006] of Google, HBase [HBase 2014] of the Hadoop project, which is an open source software, and Cassandra [Cassandra 2014], which was developed by Facebook and became an open source software later.

Generally, given key data, key value stores are suitable for searching non-key data (attribute values) associated with the key data. One of the methods will be explained below.

First, a hash function is applied to a node which stores data. According to the result of the hash function, the node is mapped to a point (i.e., logical place) on a ring type network (see Fig. 2.9).

In storing data, the same hash function is applied to a key value of each data and then the data is similarly mapped to a point on the ring. Each data is stored in the nearest node by the clockwise rotation of the ring. Although movement of data occurs depending on addition and deletion of nodes, the scope of the impact can be localized. Thus, for data access, you just need to search for the nearest node located by applying the hash function to the key value. This access structure is called consistent hashing, which is also adopted by P2P systems used for various purposes such as file sharing.

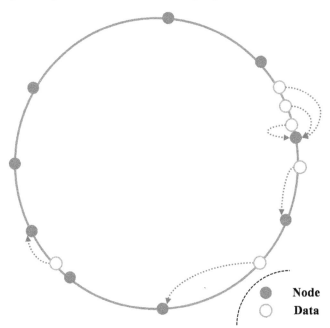

Figure 2.9 Consistent hashing.

It is inefficient for large-scale data access that data are searched from a key value store by specifying conditions on non-key data or inequality conditions on key data. That is because it is necessary to check the conditions on every data.

In order to search data efficiently by conditions on attributes other than keys, there are typical methods such as indexing attribute values. However, if indexing would be used, the kinds of indexes may increase in number and the size of the whole indexes may be dozens of times the size of the original data. Alternatively, there is a method of building keys of an index by combining values of one or more attributes. In this case, if the variation of values of each attribute is large, combinatorial explosion may be caused in turn.

Currently, relational database management systems (RDBMS) or SQL as the search interfaces are mainstream mechanisms for storing and searching large data. They provide access methods capable of efficiently performing more complicated queries than key value searches, including select and join operations (i.e., comparison by the keys of two or more tables).

On the other hand, if the user does the same operations on key value stores, a problem occurs. The users have to program the logic corresponding to complicated queries. It seems promising as a solution to this problem to build functionally-rich query processing facilities as a kind of middleware on top of key value stores. Thereby, key value stores would be apparently close to conventional database management systems as a whole. However, this might not lead to solving performance issues as to processing complex queries because the underlying key value stores are inherently poor at processing them.

Owing to the simple structures, key value stores can provide more scalability than conventional relational databases. Furthermore, key value stores have been as specialized to particular promising applications (i.e., Web services) as possible. On the other hand, relational databases have been developed as general-purpose database management systems which are aimed to flexibly and efficiently support storing, searching, and updating of various databases. It can be said that relational databases or SQL interfaces are over-engineered for the requirements of the current Web services at least in the facilities for advanced search and highly reliable data management.

2.8.2 MapReduce—a Mechanism for Parallel Distributed Computing

In general, the types of algorithms that are often reused are called design patterns. MapReduce [Dean et al. 2004] is considered as a design pattern which can process tasks efficiently by carrying out scale-out in a straightforward manner. For example, human users browsing web sites and robots aiming at crawling for search engines leave the access log data in Web servers when they access the sites. Therefore it is necessary to extract only the session (i.e., a coherent series of page accesses) by each user from the recorded access log data and store them in databases for further analysis. Generally such a task is called extraction, transformation, and loading (extract-transform-load, ETL). To extract pages, search terms and links from pages crawled by search engines and store them in repositories or to create indices by using such data, are also considered as ETL tasks.

MapReduce is suitable for applications which perform such ETL tasks. It divides a task into subtasks and processes them in a parallel distributed manner. In a word, subtasks are mapped to two or more servers so as to be processed and each result is shuffled and aggregated into a final result (see Fig. 2.10). MapReduce is suitable for cases where only data or parameters of each subtask are separate although the method of processing is completely the same. An example which calculates the frequencies of search terms in the whole page set using MapReduce is shown in Fig. 2.10. First, the Map phase is carried out and the outputs are rearranged (i.e., shuffled) so that they are suitable for the input of the Reduce phase. In other words, for applications where similarity (i.e., identity of processing in this case) and diversity (i.e., difference of data and parameters for processing) are inherent, MapReduce exploits these characters to improve the efficiency of processing.

On the other hand, there is a research report about the performance of MapReduce [Stonebraker et al. 2010]. Stonebraker of MIT executed several different tasks by using a computer cluster which consisted of 100 nodes.

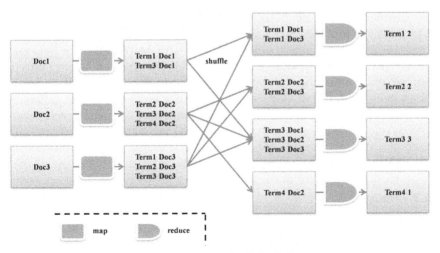

Figure 2.10 An example of MapReduce.

The performances of MapReduce on top of Hadoop and the newest parallel RDBMS (specifically, column-oriented Vertica and row-oriented DBMS-X) were compared. Parallel RDBMS outperformed MapReduce by several times in tasks such as Grep-like parallel search of a character string and Web log analysis. The former outperformed the latter by more than ten times in complicated query processing.

Responses to the fault tolerance of data are also different between MapReduce and RDBMS. If an error occurs in RDBMS, it tries to recover from the error state on the spot immediately. On the other hand, even if an error occurs, MapReduce leaves the state as it is and continues the current process. In other words, MapReduce takes the position that data have only to be consistent when data are finally needed (called eventually consistent).

CAP theorem [Brewer 2000] states that three characteristics, consistency (C), availability (A), and partition tolerance (P) cannot be fulfilled simultaneously by application systems in distributed environments. However, any two of them can be satisfied at the same time. Web Services, which think highly of availability and partition tolerance, have selected the concept of being eventually consistent as supported by MapReduce.

HBase emphasizes A and P among CAP. On the other hand, current DBMS attach more importance to C and A. According to Stonebraker, while parallel RDBMS are suitable for processing complex queries over structured data and applications whose data are frequently updated, MapReduce is suitable for the following applications.

- ETL system
- Data mining involving complicated analysis
- XML data processing

Furthermore, in comparison with RDBMS, MapReduce is a ready-to-use, powerful tool obtained at a low cost. In other words, the scenes where they are useful differ by parallel RDBMS and MapReduce. In general, ecosystems are social systems based on technology which are self-organized and are maintained through interactions among the members like natural ecosystems. Parallel RDBMS and MapReduce have evolved in different ecosystems. However, these two ecosystems may influence each other and develop into a new one in course of time like natural ecosystems. For example, a recent research of Google on the database system F1 [Shute et al. 2013] is aimed at balancing CAP on all the aspects.

2.8.3 Hadoop

Hadoop [Hadoop 2014] is an open source software for distributed processing on a computer cluster, which consists of two or more servers. Hadoop corresponds to an open source version of a distributed file system

GFS (Google File System) of Google and a distributed processing paradigm Hadoop of Google, which uses GFS. Currently, Hadoop is one of the projects of Apache.

Hadoop consists of a distributed file system called HDFS (Hadoop Distributed File System) that is equivalent to GFS, MapReduce as it is, and Hadoop Common as common libraries (see Fig. 2.11).

As MapReduce has already been explained, only HDFS will be briefly explained here. A computer system is a collection of clusters (i.e., racks in reality) which consist of two or more servers (see Fig. 2.12). Data are divided into blocks. While one block for original data is stored in a server which is determined by Hadoop, copies of the original data are stored in two other servers (default) inside racks other than the rack holding the server for the original data simultaneously.

Although such data arrangement has the objective to improve availability, it also has another objective to improve parallelism.

The special server called NameNode manages data arrangement in HDFS. The NameNode server carries out book keeping of all the metadata of data files. The metadata are resident on core memories for high speed access. Therefore, the server for NameNode should be more reliable than the other servers.

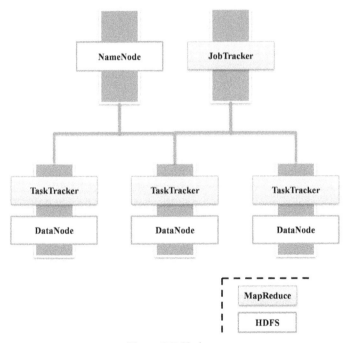

Figure 2.11 Hadoop.

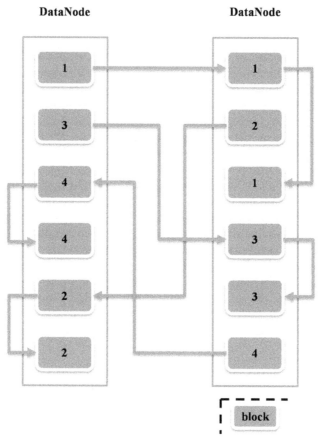

Figure 2.12 HDFS.

It is expected that if copies of the same data exist in two or more servers, candidate solutions increase in number for such problems that process tasks in parallel by dividing them into multiple subtasks. If Hadoop is fed a task, Hadoop searches the location of relevant data by consulting NameNode and sends a program for execution to the server which stores the data. This is because communication cost for sending programs is generally lower than that for sending data.

In general, programs basically perform operations on individual data. However, Hadoop basically treats a set of data. Pig and Hive, two different kinds of programming environments, have been created depending on how the concept of a set is introduced into programming. Pig is based on the concept of data flow rather than a simple set of data. Pig repeatedly performs the same operation to each element of a set of data. On the other hand, Hive performs operations on a set of data like SQL of RDBMS. That is, it can be said that Hive is a result of efforts for compensating one of the weak points of Hadoop in comparison with RDBMS as described previously. However, there is no change to the fact that Hive sequentially accesses data internally.

References

[Brewer 2000] Eric Brewer: Towards Robust Distributed Systems http://www.cs.berkeley.edu/~brewer/cs262b-2004/PODC-keynote.pdf accessed 2014
[Cassandra 2014] Cassandra http://cassandra.apache.org/accessed 2014
[Chang et al. 2006] Fay Chang, Jeffrey Dean, Sanjay Ghemawat, Wilson C. Hsieh, Deborah A. Wallach, Mike Burrows, Tushar Chandra, Andrew Fikes and Robert E. Gruber: Bigtable: A Distributed Storage System for Structured Data, Research, Google (2006).
[Dean et al. 2004] Jeffrey Dean and Sanjay Ghemawat: MapReduce: Simplified Data Processing on Large Clusters, Research, Google (2004).
[DeCandia et al. 2007] Giuseppe DeCandia, Deniz Hastorun, Madan Jampani, Gunavardhan Kakulapati, Avinash Lakshman, Alex Pilchin, Swaminathan Sivasubramanian, Peter Vosshall and Werner Vogels: Dynamo: Amazon's highly available key-value store. Proc. SOSP 2007: 205–220 (2007).
[DynamoDB 2014] DynamoDB http://aws.amazon.com/dynamodb/accessed 2014
[Hadoop 2014] Hadoop http://hadoop.apache.org/accessed 2014
[HBase 2014] HBase https://hbase.apache.org/accessed 2014
[IDC 2008] IDC: The Diverse and Exploding Digital Universe (white paper, 2008). http://www.emc.com/collateral/analyst-reports/diverse-exploding-digital-universe.pdf Accessed 2014
[IDC 2012] IDC: The Digital Universe In 2020: Big Data, Bigger Digital Shadows, and Biggest Growth in the Far East (2012).
http://www.emc.com/leadership/digital-universe/iview/index.htm accessed 2014
[Kline 2011] R.B. Kline: Principles and practice of structural equation modeling, Guilford Press (2011).
[Mcfedries 2012] P. Mcfedries: The coming data deluge [Technically Speaking], Spectrum, IEEE 48(2): 19 (2011).
[Shirai et al. 2013] Motohiro Shirai, Masaharu Hirota, Hiroshi Ishikawa and Shohei Yokoyama: A method of area of interest and shooting spot detection using geo-tagged photographs, Proc. ACM SIGSPATIAL Workshop on Computational Models of Place 2013 at ACM SIGSPATIAL GIS 2013 (2013).
[Shute et al. 2013] Jeff Shute et al.: F1: A Distributed SQL Database That Scales, Research, Google (2013).
[Stonebraker et al. 2010] Michael Stonebraker, Daniel J. Abadi, David J. DeWitt, Samuel Madden, Erik Paulson, Andrew Pavlo and Alexander Rasin: MapReduce and parallel DBMSs: friends or foes? Communication ACM 53(1): 64–71 (2010).
[Taboada et al. 2011] Maite Taboada, Julian Brooke, Milan Tofiloski, Kimberly Voll, Manfred Stede: Lexicon-Based Methods for Sentiment Analysis, MIT Computational Linguistics 37(2): 267–307 (2011).
[Vogels 2007] W. Vogels: Data Access Patterns in the Amazon.com Technology Platform (Keynote), VLDB (2007).
[Zikopoulos et al. 2011] Paul Zikopoulos and Chris Eaton: Understanding big data Analytics for Enterprise Class Hadoop and Streaming Data, McGraw-Hill (2011).

Hypotheses in the Era of Big Data

The big data era, has made it more difficult than before, for us to construct hypotheses. However, the role of hypotheses is increasingly more important. In this chapter, the essence of hypotheses in the big data age will be explained first. Then classical reasoning forms, such as induction, deduction, and analogy, will be discussed as fundamental techniques for constructing hypotheses. After that, abduction as plausible reasoning and causalities and correlations as basic concepts related to reasoning will be summarized.

3.1 What is a Hypothesis?

Relationships between big data and hypotheses will be described here. In the era of big data, it has become more important and more difficult for us to make a promising hypothesis beforehand. A hypothesis, in general, is a provisional explanation based on the observed values as to a certain phenomenon. In a narrower sense, it is a predictive relationship between variables equivalent to observable causes and variables equivalent to observable results. Moreover, a hypothesis must be verifiable. However, verification of a hypothesis is not equal to proof of the hypothesis. Even only one counter example can prove that a hypothesis is incorrect. On the other hand, it is very difficult to prove that a hypothesis be totally correct or incorrect. In other words, verification of a hypothesis is to quantitatively evaluate whether a hypothesis is acceptable or not after a related phenomenon occurs.

Is a hypothesis necessary in the first place? Indeed, without constructing a hypothesis beforehand, a certain kind of prediction may be attained by considering feature vectors of thousands of dimensions representing all the conceivable variables and feeding such vectors into machine learning or data analysis libraries runnable on Hadoop, which is a parallel software platform working on cluster computers.

A case where a hypothesis-free method was a success has been reported by genome researchers. For example, with the technique of association study applied to the whole genome, the fact that a specific gene adjusts development of symptoms of Parkinson's disease through an interaction with coffee has been discovered [Hamza et al. 2011].

However, the prediction obtained by only the above method cannot explain any mechanism or validation for the prediction. If such an explanation is difficult, it will not be so easy for the users to adopt this mechanism as a certain prediction with confidence. Thus, it is definitely better to construct a hypothesis beforehand in data analysis or data mining even if it is the big data era now.

In general, hypotheses also have life cycles (see Fig. 3.1). After a problem is defined, a hypothesis is constructed through pre-analysis of the problem (hypothesis construction). After data relevant to the constructed hypothesis are collected, some methods such as statistical analysis and experiments are applied to the data, thereby verifying the hypothesis (hypothesis verification). If the hypothesis is accepted as the result of verification, then it is, so to say, promoted to a theory (theory formation). Otherwise, the rejected hypothesis is either thrown away or corrected for another life cycle.

Here a position of the hypothesis in data mining will be explained. Simply put, data mining consists of tasks for generating a model (i.e., pattern) equivalent to a hypothesis, based on collected data. The techniques used in data mining are often called machine learning. Classification is a supervised technique which uses parts of samples as training data for constructing a hypothesis and other parts of samples as testing data for validating the hypothesis, whose class (i.e., classification category) is known

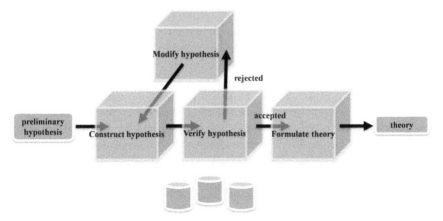

Figure 3.1 Life cycle of a hypothesis.

in advance. Association rule analysis and cluster analysis as unsupervised techniques can be performed without using training data or constructing any hypothesis in advance.

However, examples of hypotheses constructed by the users in advance of each task of data mining or hints provided by the users to data mining tasks for hypothesis construction may be a great help by guiding the selection of suitable techniques or parameters for obtaining a better model as a target in data mining. For example, it can be said that learning classification rules or other models after narrowing down beforehand to attributes which the users consider as important for classification is equivalent to giving some hints to the data mining task for hypothesis construction. As to association rules, the interest level of the users as to kinds, support values, and confidence values of frequent items can be conjectured by analyzing frequent item sets (or association rules) which the users construct based on their experiences. Furthermore, in clustering, to specify constraints that a set of data finally belong to the same cluster or to specify constraints that they finally belong to separate clusters is equivalent to giving a certain kind of guideline to hypothesis construction.

In data mining, of course, whether a hypothesis (that is, a model) is acceptable or not is mainly judged by the measure of accuracy. Simultaneously, the value of a hypothesis is evaluated by the interest level, so to speak, interestingness for the users in a certain field. It can be stated that a hypothesis prior to data mining or a hint given to data mining reflects expressed interest level in the field in a certain sense. Therefore, it can be said that hypotheses and hints which are provided by the users in advance are also effective so as to measure the interest level of the hypothesis mechanically made by data mining. Domain specialists' empirical rules and expectations often precede with analysis by using even unsupervised techniques such as association rule analysis and cluster analysis in data mining. This book will mention cases where more valuable discovery can be made by analyzing social data by focusing on people's interests such as shooting directions in photos and terms cited in microblogs.

Now hypotheses in science will be lightly referenced. Even scientific discoveries which seem to be accidental require construction of a hypothesis based on careful observation of related phenomena, sharp intuition (or inspiration), and deep discernment as well as elaborate experiments to verify the hypothesis. An old hypothesis will be replaced by a new hypothesis if the new one can explain the phenomenon more accurately than the old one. For example, as the world of particles constituting materials has been unraveled, the theory explaining the interactions among the particles has changed from Newtonian mechanics to quantum mechanics.

If a phenomenon which cannot be explained by existing hypotheses emerges, then the hypotheses will be rejected. In this way, verified hypotheses make human beings' scientific knowledge richer than ever. Indeed, there are some opinions that scientific research can be done only by data intensive computing [Hey 2009]. Such opinions are not necessarily true in all the fields of science. That is, science essentially has a characteristic of being led by hypotheses (i.e., hypothesis-driven).

In general, a hypothesis must satisfy the following characteristics:

- A hypothesis must be able to explain as many previous examples of phenomena as possible (universal). This is what must be considered in the first place in order to construct a hypothesis.
- A hypothesis must be made as simple as possible (parsimonious). In other words, a hypothesis must be intelligible to the users.
- A hypothesis must be able to predict what will happen in the future as to a phenomenon (predictive). This characteristic is related with usefulness of a hypothesis.
- A hypothesis must be verifiable (testable).

In this book the author would like to add the following characteristic to the above list.

- A hypothesis, whether science- or business-oriented, must reflect interests of stakeholders in a certain field of the contemporary time.

3.2 Sampling for Hypothesis Construction

As this is the era of big data, it may be good news for data mining, in a certain respect, that a sufficient number of samples can be obtained more easily than previously. This situation will be explained in more detail.

In order to plan hypothesis verification, the following procedure has been followed by data analysts, a long time before the big data era.

The whole sample data, whose categories are usually given in advance by humans, are divided into one portion (i.e., training set) used so as to construct a hypothesis and another portion (i.e., testing set or validation set) used so as to measure the accuracy of the hypothesis. Then a hypothesis is constructed based on the training set and the accuracy of the hypothesis is measured with the testing set.

However, if the total number of sample data is so small, enough data for hypothesis construction cannot fully be obtained. Furthermore, it may be highly probable that the accuracy of a hypothesis will be excessively influenced by a small number of samples.

In order to solve this problem, a method called k-fold cross validation method has been invented. In the k-fold cross validation method, sample data are divided into k pieces of data in advance. Then, (k-1) pieces are used for construction of a hypothesis and the remaining one piece is used for verification of the hypothesis. A pass of experiment (i.e., data analysis or data mining) is conducted k times by exchanging the data used for hypothesis construction and computing an accuracy for each pass. The average of accuracy is computed in order to verify the hypothesis.

As the number of data naturally increases in the era of big data, it will become easier to obtain data whose size is sufficient for hypothesis construction. That is, sample big data are simply divided into two and then one of them is used for construction of a hypothesis and the remainder is used for verification of the hypothesis. And what is necessary is just to exchange these roles for the next pass and to perform data analysis or data mining once again. That is, theoretically speaking, it is always possible to perform the significant 2-fold cross validation method.

In the big data era, it may still be necessary to prepare data whose categories are known beforehand as a training set and a validation set in supervised leaning such as classification. In such a case, the data set should be prepared usually by human help. However, in the big data era, it becomes quite difficult to do such work due to the size of data. Of course, there exist solutions for such a problem [Sheng et al. 2008]. That is, it may be sometimes effective to use human power tactics like Mechanical Turk of Amazon [Amazon Mechanical Turk 2014].

3.3 Hypothesis Verification

As mentioned earlier, the number of available samples have increased since we have entered into the era of big data. However, the era of big data does not necessarily bring only good news.

Hypothesis verification will be considered here. A hypothesis is verified as follows (for example, T-test).

1. The opposite of a hypothesis to verify called an alternative hypothesis is constructed as a null hypothesis.
2. The significance level alpha is determined.
3. Based on sample data, a statistic value and p value (stated below) are calculated.
4. If the p value is smaller than the specified significance level, there is a significant difference. As a result, the null hypothesis is rejected and the alternative hypothesis is accepted.
5. Otherwise, it is judged that there is no significant difference between the two hypotheses.

Assuming that a null hypothesis is right, the p value expresses a probability (significance probability) that a certain statistic becomes more extreme than the value which is calculated based on sample data. The p value can be calculated from a test statistic (for example, T value in a T-test). On the other hand, the significance level is the tolerance limit of the probability which rejects a null hypothesis accidentally when it is right. The significance level is a criterion of judgment which the analyst can choose among values such as 0.1, 0.05, and 0.01.

We will focus on the p value here. The p value tends to decrease with an increase in the number of samples. That is, if the number of samples increases, it will be possible to lower the significance level accordingly. In other words, an alternative hypothesis becomes more probable in a purely statistical sense. This can also be said about a correlation coefficient. That is, the significance level can be decreased as much as the number of data increases, assuming that the value of a correlation coefficient is the same.

A hypothetical significance level can be decreased only because in the big data era the amount of data has increased. However, it cannot be overemphasized that there is no guarantee that more important and more interesting hypotheses can be automatically constructed. In short, the degree of interests needs to be explicitly expressed as a certain kind of hypothesis made in advance of analysis.

In addition, instead of the p value, an effect size [Kline 2011] such as Cohen's *d*, which cannot be easily influenced by the number of samples, are used nowadays.

3.4 Hypothesis Construction

In this section, hints for constructing hypotheses will be intuitively described from a viewpoint of various kinds of reasoning and analysis of causal relationships.

In general, the reasoning forms of propositions in mathematics include induction and deduction. However, please note that propositions described here are not confined to such mathematical ones, which are universally true, that is, true anywhere and anytime. It is well known that the interest level for businesses may become lower in case of too general propositions or common propositions. This book also treats propositions which are true dependent on situations, in other words, not always true, i.e., true with a certain confidence (e.g., probability). Not only mathematically strict reasoning such as induction, deduction, but also plausible reasoning and analogy are necessary, depending on realistic applications.

From a viewpoint of constructing hypotheses, these various reasoning forms will be explained below.

3.4.1 Induction

We ordinarily create propositions by generalizing individual observations (samples) o_i. This form of reasoning is called inductive inference. A simple case of inductive inference is shown below.

   $p(o_1)$ and $p(o_2)$ and ... $p(o_n)$ (i.e., if all hold)
   ----- (i.e., then the following is derived)
   $p(o)$ (i.e., universally holds)

Here o denotes a variable over each element of a set of observable data, represented by o_i, and $p(o)$ states that the variable o satisfies the proposition p.

For example, let's consider a mathematical proposition about integers [Polya 2004]. It can be expressed as follows:

$$1^3 = \left(\tfrac{1 \cdot 2}{2}\right)^2 \qquad\qquad\qquad p(1)$$

$$1^3 + 2^3 = \left(\tfrac{2 \cdot 3}{2}\right)^2 \qquad\qquad\qquad p(2)$$

$$1^3 + 2^3 + 3^3 = \left(\tfrac{3 \cdot 4}{2}\right)^2 \qquad\qquad\qquad p(3)$$

...

$$1^3 + 2^3 + 3^3 + ... + n^3 = \left(\tfrac{n \cdot (n+1)}{2}\right)^2 \qquad\qquad\qquad p(n)$$

If this proposition $p(n)$ is put in terms of geometry, it claims that the sum of the volume of the cube whose side is i ($i = 1$ to n) on the left hand of the equation is equal to the area of a square whose side is the sum of i ($i = 1$ to n) on the right hand (see Fig. 3.2). However, this is an example of

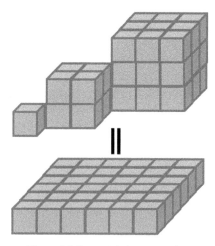

Figure 3.2 Geometric interpretation.

an ideal hypothesis based on induction, which can be proved exactly by mathematical induction.

Induction is a reasoning form which can be directly used in order to construct general hypotheses from individual observations. As a matter of course, induction-based hypotheses can explain existing observation data. Furthermore, it is important for such hypotheses to be able to explain newly coming data, too.

3.4.2 Deduction

Is deduction [Jaynes 2003], or deductive inference, which is frequently used in mathematics, also applicable to hypothesis construction? Generally in strict deductive inference, the following proposition is considered.

$P \Rightarrow q$

Here $p \Rightarrow q$ means that the proposition q is true, supposing the proposition p is true. Thus, it is said that p implies q. When the upper proposition holds, if p is right, it can be inferred that q is right. This can be expressed as follows.

$(p \Rightarrow q)$ and p true

-----

q true

Now a proposition about climates will be considered as an example. The following A and B are considered as concrete propositions p and q, respectively.

Let $A \equiv$ {El Niño/La Niña-Southern Oscillation occurs} and
$B \equiv$ {It becomes a warm winter in Japan}

Thus, if El Niño/La Niña-Southern Oscillation occurs, then it becomes a warm winter in Japan.

The above reasoning is called syllogism. When a hypothesis is constructed based on deductive inference, it is ideal to prove that a new hypothesis is right by combining already accepted hypotheses (i.e., theories). Indeed, in mathematics this is always possible. In other fields, however, hypotheses generated by deductive inference as well as those generated by inductive inference must undergo the verification process using real data.

If a hypothesis can be translated into another form (for example, simulation program) by which the hypothesis can be realized, there is also a method of comparing the result computed by the simulation program with the actually observed data and of verifying the hypothesis indirectly. Simulation programs are also used for prediction of future results along this method.

Now, generally in deductive inference, the counter-proposition $q \Rightarrow p$ is not necessarily true just because the proposition $p \Rightarrow q$ is true. That is, although the following proposition is true and q is true as well, in strict deductive inference it cannot be reasoned that p is true. Generally, the following reasoning that is made by transforming the above-described syllogism doesn't hold.

$(p \Rightarrow q)$ and q true

-----

p true

Assume that A as p and B as q are the same as those in the above example. Even if it is true that it became a warm winter in Japan (B), it cannot be claimed that the El Niño/La Niña-Southern Oscillation took place (A).

3.4.3 Plausible Reasoning

On the other hand, it may be sometimes thought that a proposition is plausible (that is, quantitatively true) in a certain field in the physical real world. In other words, it corresponds to giving quantitative reliability or plausibility to a proposition. Here plausible reasoning [Jaynes 2003] will be explained.

The following syllogism will be considered again assuming that A and B are the same as those in the previous example of proposition.

$(A \Rightarrow B)$ and A true

B true

Here assuming that the reliability of the above proposition is expressed by the following conditional probability [Jaynes 2003], for example, let us continue our discussion.

$P(B \mid (A \Rightarrow B) A)$

In order to decide the value of the reliability, the following tautology about conditional probability (aka Bayes' theorem), which is always true, will be considered first.

$P(XY \mid Z) = P(X \mid YZ)P(Y \mid Z) = P(Y \mid XZ)P(X \mid Z)$

Letting $X = A$, $Y = B$, and $Z = (A \Rightarrow B)$ in this formula, the formula can be transformed as follows.

$P(B \mid (A \Rightarrow B) A) = P(B \mid A \Rightarrow B)P(A \mid B (A \Rightarrow B))/P(A \mid A \Rightarrow B)$

In this case, of course, the value of this expression is equal to 1. It is because the denominator of the above formula can be generally transformed as follows and the underlined part in the new formula is equal to 0 in this case.

$$P(A \mid A \Rightarrow B) = P(A(B+\neg B) \mid A \Rightarrow B) = P(AB \mid A \Rightarrow B) + P(A\neg B \mid A \Rightarrow B)$$
$$= P(B \mid A \Rightarrow B)P(A \mid B (A \Rightarrow B)) + P(\neg B \mid A \Rightarrow B)\underline{P(A \mid \neg B (A \Rightarrow B))}$$

Next, the following case that is modified from the syllogism will be considered again

$(A \Rightarrow B)$ and B true

A true

Of course, this is incorrect in strict deductive reasoning. However, plausible reasoning concludes that A is more plausible as follows.

$(A \Rightarrow B)$ and B true

A more plausible

The confidence at that time is given by the value of the probability $P(A \mid (A \Rightarrow B) B)$. In order to decide the value, the above tautology (Bayes' theorem) will be used again as follows.

$$P(A \mid B (A \Rightarrow B)) = P(A \mid A \Rightarrow B)P(B \mid A (A \Rightarrow B))/P(B \mid A \Rightarrow B)$$

When the denominator of the above formula is focused on, the smaller the probability that B holds under the hypothetical $A \Rightarrow B$, the larger the probability that A holds. In other words, if a rare event (i.e., B) occurs, the plausibility of A increases. Since $P(B \mid A (A \Rightarrow B)) = 1$ and $P(B \mid A \Rightarrow B)$ <= 1, the above formula gives the following:

$$P(A \mid B (A \Rightarrow B)) >= P(A \mid A \Rightarrow B)$$

Similarly the syllogism can be extended as follows:

$(A \Rightarrow B)$ and A false

B less plausible

That is, in this case if it turns out that A is false under the hypothesis, it can be interpreted as the plausibility of B decreases (i.e., less plausible). Similarly, the confidence $P(B \mid \neg A (A \Rightarrow B))$ is calculated to be equal to $P(B \mid A \Rightarrow B)P(\neg A \mid B (A \Rightarrow B))/P(\neg A \mid A \Rightarrow B)$ by using Bayes' theorem. Further, $P(B \mid \neg A (A \Rightarrow B))$ <= $P(B \mid A \Rightarrow B)$ can be shown.

3.4.4 Abduction

If B is observed under the hypothesis that A ⇒ B, it may be thought that A can be presumed as one of the most plausible causes. This is a kind of plausible reasoning described above. Such a presumption is called abduction [abduction-SEP 2014]. In abduction, the capability to presume is important. Therefore, the following probability will be considered by using Bayes' theorem.

$$P(A \Rightarrow B | B) = P(A \Rightarrow B)P(B | A \Rightarrow B)/P(B)$$

The power of reasoning as to presumption can be measured quantitatively by this probability. That is, in general, when the result (i.e., evidence) B is observed, all that is necessary is to presume, as a cause of B, A such that A makes the above probability the highest among all the considered hypotheses.

Thus, when creating a hypothesis for the purpose of presumption, it is sometimes worth considering to use a certain proposition for the reverse direction. This is important, especially for construction of hypotheses in exploratory data analysis.

3.4.5 Correlation

Assuming that the phenomena C1 and C2 often occur in a certain field either simultaneously or irrespective of chronological sequences, if either C1 or C2 occurs, it will be thought that the rest may occur. In such a case, there may be some correlations [Kline 2011] between C1 and C2. That is, in such a case (i.e., high co-occurrence), both a hypothesis and its contrary are considered simultaneously for the time being.

$$C_1 \Rightarrow C_2$$
$$C_2 \Rightarrow C_1$$

For example, let C1 = {inductive} and C2 = {reasoning}. Then it is expected that this pair will be used often together in documents or Web pages. However, it is not sufficient for analysis of correlations to observe only the frequency of the above co-occurrence. If C1 (or C2) generally increases as C2 (or C1) increases, it is said that there is a positive correlation between C1 and C2. On the other hand if C1 (or C2) occurs more often, then C2 (or C1) occurs less often, depending on cases. In such a case, it is said that there is a negative correlation between them. Moreover, it is said that any correlation is absent when there is neither a positive correlation nor a negative correlation. If there are a lot of co-occurrences but no correlations, it means that just coincidences occurred.

The check of a correlation is usually performed by calculating a correlation coefficient such as cosine measure and Jaccard coefficient, as will be described later. A positive correlation, a negative correlation, and no correlation are judged with the value of the coefficient.

For example, C_i are assumed to take the following values:

C_1 = {inductive}, C_2 = {reasoning}, C_3 = {deductive}

A positive correlation is likely to be between C1 and C2 (i.e., {inductive reasoning}) and between C3 and C2 (i.e., {deductive reasoning}). On the other hand, although there are also many co-occurrences between C1 and C3 (i.e., {inductive deductive}), it may be expected that there is a negative correlation.

3.4.6 Causality

Then, in relation to the already mentioned deductive inference, let us consider a case where there is a strict causal relationship such that C is brought about as a result of B. In this book, such a case is also expressed as follows.

$B \Rightarrow C$

For example, propositions B and C are considered as follows (see Fig. 3.3).

$B \equiv$ {It is a warm winter in Japan}
$C \equiv$ {The sales of winter clothing falls in Japan}

That is, if it is a warm winter in Japan (B), then the sales of winter clothing falls in Japan (C).

In pure science, it is not claimed that there is any causal relationship between B and C even if there is any correlation between B and C, or even if they occur simultaneously or continuously. Indeed, although it is rare, both B and C may occur by complete chance.

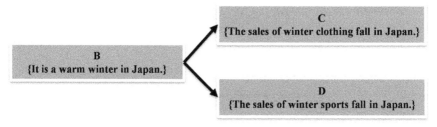

Figure 3.3 Common cause.

Moreover, one cause may produce two or more results. In such a case, if put in another way, the cause B is a common cause of C and D. This case can be described as follows.

B ⇒ C
B ⇒ D

That is, although there is apparently a positive correlation between C and D in this case, there is no direct dependency between C and D. The relationship in that case is called spurious correlation (i.e., pseudo correlation) in distinction from true correlation.

For example, B and C of the above example are used as they are and D is set as follows (see Fig. 3.4).

D ≡ {the sales of winter sports fall in Japan}
Both B ⇒ C and B ⇒ D hold.

Although there seems to be a positive correlation between the sale of winter clothing (C) and the sale of winter sports (D), there is no clear direct dependency.

When the cause B is not known yet or B is inherently latent (that is, B cannot be observed), therefore, when a correlation of B and C and a correlation of B and D cannot be directly measured, it is sometimes considered effective to use the relationship between C and D conditionally and partially.

Furthermore, the phenomenon M exists between A and C and M is both a result of A and a cause of C. It is generally expressed as follows:

A ⇒ M ⇒ C

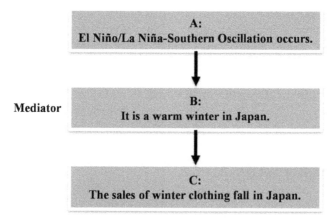

Figure 3.4 Mediator.

In this case there exist strict causal relationships between A and M as well as between M and C. Furthermore, A influences C through M. The last condition is especially important. Generally M is called mediator in this case. The role of a mediator is called mediation. And it is said that the influence which A has on C is indirect.

The previous A, B, and C are used as they are and let M = B. If El Niño/ La Niña-Southern Oscillation occurs (A), it will be a warm winter in Japan (B) as the result of (A) and then the sales of winter clothing falls down (C) owing to the result (B).

Please note that A may influence C both indirectly and directly depending on applications.

Moreover, two or more causes may be related to one result. In the case where it is possible that all the causes are related to the result, multiple linear regression analysis or more general techniques such as SEM (Structural Equation Modeling) [Kline 2011] are useful. Furthermore, a part of causal relationship (i.e., precondition) may be described by combining the conditional expressions associated with individual variables by logical operators (i.e., And, Or, Not).

Here a causal relationship will be defined strictly (i.e., classically in a certain sense).

It is said that there exists a causal relationship (A ⇒ B) between A and B if the following conditions are satisfied:

(Condition one) A precedes B temporarily.
(Condition two) There is any correlation between A and B.
(Condition three) There is no mediator between A and B.
(Condition four) B does not occur without A.
(Condition five) A relationship A ⇒ B holds universally.

Some remarks will be made about each of the above conditions of a causal relationship from a modern viewpoint.

In the condition one, time lags or time scales in a case where A happens in advance of B are different, depending on application domains. In natural phenomena, it may be shorter than one second or it may be measured in the unit of 100 years. For example, El Niño/La Niña-Southern Oscillation affects a warm winter of Japan in over half a year.

An interaction with a feedback loop such that A causes B and, in turn, B causes A can be also considered. In that case, when the interaction is viewed as a whole, it seems that A and B occur simultaneously. Also in a case where A is temporarily unchanged or stabilized, it may be thought that A agrees on the condition of precedence.

In the condition two, a true correlation but not a spurious correlation is considered. Therefore, B is dependent on A.

Associated with the condition three, it may sometimes also be necessary to consider causal structures including transitive causal relationships, that is, a combination of strict causal relationships described here. For example, $A \Rightarrow B \Rightarrow C$ will be considered. If the effect $A \Rightarrow C$ is also accepted as direct when B works as a mediator between A and C, the causal structure can be considered as a synthetic causal relationship consisting of both direct and indirect effects (i.e., transitive effect).

The condition four eliminates a case where A does not occur and B occurs. In such a case another cause U can be considered to lead to the result B. A so-called counterfactual dependency needs to be checked in the case. That is, it is necessary to verify that B does not occur if A does not occur. The statistical method which makes two groups from sample data selected at random, lets one group and another be A and non-A, respectively, and verifies a hypothesis by experiments using the two groups, is a kind of test method of this counterfactual dependency. If U has caused B, there is a strict causal relationship between U and B. Of course, if there are two or more causes, each contribution must be considered quantitatively.

In the condition five, from a viewpoint of usefulness of a causal relationship, it is necessary to redefine the meaning of being universal, depending on applications. In that case, it is equivalent to considering, so to speak, micro universality as a situation where some conditions are given on time and space. For example, B claims that it becomes a warm winter not in the whole world but only in Japan.

3.4.7 Analogy

In this subsection, analogy [Polya 1990] will be explained as one form of reasoning. In analogical reasoning, let's consider a case where a proposition T to create is structurally similar to an existing proposition S. In comparison of the portions which correspond to each other in T and S, what is not yet known in T but is already known in S can be predicted for the very reason. To consider a proposition according to this process is called analogy.

For example, hypotheses about forces in physics will be considered. According to particle physics, forces (or interactions) between elementary particles are classified into four kinds: electromagnetic force, weak force, strong force, and gravity.

It has already been verified that the first three kinds of forces, that is, electromagnetic, weak, and strong forces, are intermediated by existing particles. Then, a new hypothesis will be produced as follows:

Electromagnetic force: Photon
Weak force: W, Z bosons
Strong force: Gluon

Gravity: Hypothetical boson called graviton (hypothesis)

That is, the four forces have similar structures; particles and relationships (i.e., interactions) between them constitute components of hypotheses on forces. In gravity, however, particles as components have not yet been known until now. As in the other three forces, particles, tentatively called gravitons, are predicted to exist in gravity as its components and to realize gravity as mediators. Some theories speculate that gravitons are not particles but strings.

Incidentally, by analyzing the results of proton-proton collision experiments which had been done for more than one quadrillion times using the particle accelerator in CERN, the existence of a Higgs boson, which has long been predicted to be a particle giving mass to everything in the Universe, has been confirmed with very high probability [Cho 2012]. As the existence of a Higgs boson was validated by more experimental data after that, two physicists Higgs and Englert who had independently predicted its existence were awarded the Nobel prize in physics in 2013. This occurrence will definitely remain forever as one of the successful examples of science big data applications.

Furthermore, although it is not analogy itself, the method of constructing hypotheses based on analogical reasoning can be considered. That is, it is to create a new proposition by either generalizing or specializing parts of propositions which have already been confirmed correct in a certain field.

3.4.8 Transitive Law

Furthermore, the following transitive law used by deductive reasoning can also be used to construct a hypothesis.

$(A \Rightarrow B)$ and $(B \Rightarrow C)$

$A \Rightarrow C$

Here, however, the transitive law is not used for the proof of a proposition. Instead, a new hypothesis is constructed by applying the transitive law to two or more hypotheses which have already been constructed. Moreover, the number of hypotheses can be increased by replacing a part or all of the propositions as the premises with the already explained plausible propositions and by applying this transitive law to them. For example, in case it is assumed that any indirect effect exists between two variables, addition of a hypothesis for verification of any possible direct effect between the variables is an example of applications of the transitive law.

Various reasoning forms explained in this chapter are applicable to cases where a new hypothesis is constructed either based on the observed data or by transforming an existing hypothesis. Moreover, plausible reasoning as well as reasoning based on strict syllogism can be used in such hypothesis construction. Which forms of reasoning should be used depends on specific techniques used for analysis according to the purpose. Regardless of which analytical methods to use, it may be possible to construct a hypothesis based on relationships between big objects (i.e., influence relationships between concepts), which are extensions of causal relationship and correlation between variables.

3.5 Granularity of a Hypothesis

In this subsection, granularity of a hypothesis, or its level of abstraction will be described.

Hypotheses in multivariate and data mining are mainly based on relationships between variables of data. They are usually hypotheses within the same big objects, or the same big data source. On the other hand, big data applications generally involve two or more big data sources. Therefore, hypotheses related to the whole applications are based on relationships between heterogeneous big objects.

Construction of hypotheses within the same big object is usually based on correlations between variables of data contained therein. In that sense, hypotheses involving the same big object, that is, intra-object hypotheses are called micro hypotheses. On the other hand, construction of hypotheses involving heterogeneous big objects is based on correlations between numbers of data (i.e., aggregative counts) rather than those between individual variables of data. In that sense, inter-object hypotheses are called macro hypotheses. Macro hypotheses are naturally more abstract than micro hypotheses.

In construction of macro hypotheses, it is the first consideration to discover a set of heterogeneous big data sources as stakeholders of correlations. Tasks for discovering mutually related data sources are based on similarities between a set of data included by heterogeneous data sources. Candidate attributes required for calculating such similarities are common attributes such as time, place, meaning, which can universally relate heterogeneous data to each other. To discover relationships between heterogeneous big data sources requires to retrieve one data source from the other or retrieve all of them simultaneously by using such attributes as universal keys. If all heterogeneous data sources contain semantics, it is sometimes effective for discovery of relationships among them to first cluster data within each of the data sources and second to cluster the results spanning over the heterogeneous data sources based on the semantics.

Please note that join keys in relational databases and object identifiers in object-oriented databases are available in addition to such universal key attributes in some big data applications. However, even if such join predicates can be logically described, it is not always possible to efficiently execute the corresponding operations on big data.

Let us consider a big data application which is based on integrated analysis of physical real world data (e.g., traffic data) and social data (e.g., Twitter articles). In this case, physical real world data lack semantic information while social data contain semantic information. First, an anomaly as an interest (e.g., a cause of traffic congestion in subways) is discovered by observing bursts or outliers in the physical real world data as one big data source (e.g., the number of passengers at a station). Next, social data (e.g., Tweets) as another big data source are retrieved by using information common to heterogeneous big data sources, such as time and place, that is, universal join keys. Text mining is applied to the retrieved data and the analyst's interest (e.g., a concert by a popular idol group as a cause of congestions) is uncovered as the result. Then causal relationships between events and congestions are more precisely and quantitatively analyzed. As a possible scenario based on the analysis result, recommendation or optimization in a broad sense (e.g., measures for easing congestions) is performed by monitoring coming similar event data (e.g., scheduled concerts by the same group).

Please note that relationships between heterogeneous social data such as Twitter and Flickr can be discovered by clustering them based on semantic similarities because both contain semantics.

Construction and analysis of micro hypotheses require data mining and multivariate analysis. On the other hand, construction and analysis of macro hypotheses require an integrated framework for describing and analyzing big data applications, especially, the big object model as the main function of the framework, as proposed in this book.

3.6 Hypothesis Revisited

Again, the author cannot help but hesitate to analyze without constructing any hypothesis beforehand. Here the author would like to think about hypotheses.

Data analysis is often performed by the following two steps of construction and confirmation of hypotheses:

(Step one) Assuming that relationships exist between variables systematically or comprehensively, a hypothesis is created, analyzed and confirmed with collected data.

(Step two) If spurious relationships are discovered in the hypothesis based on the result of the first step, they are deleted. Then a new hypothesis is created and confirmed again by using the collected data although it is not necessary to collect data anew in principle.

Here, the hypothesis created in the first step is called a weak hypothesis and the hypothesis created in the second step is called a strong hypothesis. Of course, both are the same in the structures or meanings as hypotheses.

The weak hypothesis and the strong hypothesis correspond to the hypothesis in exploratory data analysis and the hypothesis in confirmatory data analysis, respectively. And the computing cost required for confirmation of the weak hypothesis is clearly higher than the cost required for confirmation of the strong hypothesis. Generally, the difference tends to further expand in big data.

Software platforms such as Hadoop, which work on top of hardware platforms such as parallel computer clusters, may reduce the overall computational cost or processing time. Therefore, in the big data era, there exist justifications for carrying out the above two steps of construction and confirmation of hypotheses.

Finally some notes about data collection will be made. In the big data era, business data and science data are already collected and accumulated in many cases. Social data, especially can be collected by crawling or using the Web Service API to some extent from the Web sites not only at generation time but after that as well. However, substantial cautions and efforts are required for collecting relevant data exhaustively.

On the other hand, physical real world data, such as everyday occurrences, volatilize in many cases unless they are consciously recorded by the application side. That is, it is of course necessary for the information systems to store such physical real world data outside the systems by a certain means.

In other words, the positioning of hypotheses has largely changed since we entered into the era of big data. In the pre-big data era, a hypothesis was constructed in the first place. After that experiments and observations were done to collect data necessary to confirm the hypothesis. On the other hand, it is currently necessary to choose, cleanse, and transform appropriate portions among the stored data prior to analysis. Such tasks correspond to data collection truly required in the era of big data. These will raise both the quality of data and that of the constructed hypothesis as a result. Hypotheses created in advance by analysts or interests (i.e., premature hypotheses) extracted from end-users by analysts helps the analysts to appropriately select data collected beforehand.

References

[Abduction-SEP] Abduction-SEP (Stanford Encyclopedia of Philosophy) http://plato.stanford.edu/entries/abduction/accessed 2014

[Amazon Mechanical Turk 2014] Amazon Mechanical Turk: Artificial Intelligence https://www.mturk.com/mturk/welcome accessed 2014

[Cho 2012] A. Cho: Higgs Boson Makes Its Debut After Decades-Long Search, Science 337(6091): 141–143 (2012). DOI:10.1126/science.337.6091.141

[Hamza et al. 2011] T.H. Hamza et al.: Genome-Wide Gene-Environment Study Identifies Glutamate Receptor Gene GRIN2A as a Parkinson's Disease Modifier Gene via Interaction with Coffee. PLoS Genet 7(8): e1002237 (2011). doi:10.1371/journal.pgen.1002237

[Hey 2009] T. Hey: The Fourth Paradigm: Data-Intensive Scientific Discovery, Microsoft Press (2009).

[Jaynes 2003] E.T. Jaynes: Probability Theory: The Logic of Science, Cambridge University Press (2003).

[Kline 2011] R.B. Kline: Principles and practice of structural equation modeling, Guilford Press (2011).

[Polya 1990] G. Polya: Induction and Analogy in Mathematics (Mathematics and Plausible Reasoning, vol. 1), Princeton University Press (1990).

[Polya 2004] G. Polya: How to Solve It: A New Aspect of Mathematical Method, Princeton University Press (2004).

[Sheng et al. 2008] V.S. Sheng, F. Provost and P.G. Ipeirotis: Get another label? improving data quality and data mining using multiple, noisy labelers, Proc. the 14th ACM SIGKDD international conference on Knowledge discovery and data mining, pp. 614–622 (2008).

Social Big Data Applications

In this chapter, the differences between social media and the general Web will be clarified first, from a viewpoint of interactions. Then various types of social big data applications that seem promising from the viewpoint of enterprise use (i.e., business use) rather than from the viewpoint of personal use will be described based on the characteristics. Furthermore, examples of hypotheses are described by using the MiPS model introduced previously as well as the analysis scenarios and required tasks.

4.1 Differences between Generic Web and Social Media as Subjects of Analysis

Before discussing the applications of social media, the differences between social media and generic Web (i.e., surface Web, but not deep Web) will be discussed by focusing on the natures of the respective users.

As discussed below, the types of user interactions with the systems are very different in the generic Web and social media. In generic Web and social media, the types of data that can be used in the analysis also differ depending on the differences of the types of such interactions.

First, the interactions in generic Web are considered. The users of generic Web are roughly classified into end users and site administrators. Creation, modification, and deletion as to the contents (i.e., pages and links) are explicitly done by the administrator on the generic Web. On the other hand, the end user interactions are mainly browsing the Web pages and the users' click streams are recorded in the Web site as user access histories. Indeed, in some sites other actions such as input of search terms can be done through some kinds of forms. However, the Web sites that demand the user accounts or allow the users to query the backend databases are the deep

Web, not the surface Web. In a word, the end users are anonymous because they cannot be identified only from the IP address. Therefore, pages and links inside and outside the sites as explicit relationships and the users' click streams as implicit relationships are important as the subject of the analysis on the generic Web. Only the site administrator can use the access histories basically.

On the other hand, there exist explicit users who can be identified by the account names in addition to the administrators in social media. The profiles of the users can also be accessed by other users. The users can perform various interactions with social media sites, which include browsing and creation of social data. Primary contents such as articles and photos, secondary contents for the primary contents such as tags or evaluations, and access histories as results of such interactions constitute social data. In course of time, the relationships between the users, those between the intra-site contents, those between the inter-site contents, and those between the users and contents are created. In addition to the contents directly or indirectly created by the users, the diverse histories and relationships created in this way are important subjects of analysis in social media. Social media are different from generic Web in that most of these data are available through the Web services API provided by social media sites.

Paying attention to interactions between social media systems and users, a typical configuration of such systems is drawn in Fig. 4.1. For the sake of comparison, a configuration of generic Web is also illustrated in Fig. 4.2.

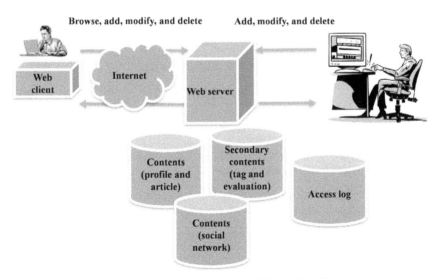

Figure 4.1 Interaction of a user with social media.

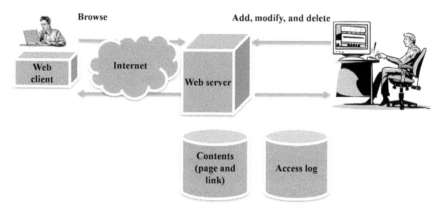

Figure 4.2 Interaction of a user with a generic Web site.

4.2 Classification of Social Media Applications based on the Components

Generally, Web mining is classified into content mining, structure mining, and usage mining by focusing on pages, links, and access histories, respectively. Similarly, various applications (tasks) of social media could be classified depending on which are observed among the contents, structures (relationships), and access histories as components of social media. That is, enterprise applications of social media can be roughly classified into the following three categories depending on which components of social media are observed by the analysts.

(1) Applications based on the analysis of contents

Applications in this category include the following:

- Analysis of sentiments and reputation of the users about certain products and services.
- Discovery of complaints and improvement ideas of the users about certain products and services.
- Investigation of wishes and needs of the users about products and services which do not exist as yet.

Furthermore, if the user profiles can be used as well, at the time of analysis, more detailed analysis will be possible.

(2) Applications based on the analysis of structures

The structures include relationships between contents, those between users, and those between both of them. Some applications discover pieces of information from social media and utilize them, focusing on one or any combination of these three relationships. Such applications include the following:

- Discovery and use of similar contents
- Discovery and use of communities consisting of similar users
- Discovery and use of users with influence

Advertisements and marketing such as recommendation of products or services can be considered as applications based on the use of the discovered contents, communities, and users. In discovery of similar contents, not only the features of the contents themselves, but also the similarity of the users of the contents should be observed. Use of user profiles may also raise the accuracy of discovery and the quality of use in these applications.

(3) Applications based on the analysis of access histories and changes

Applications based on the dynamic analysis of contents or users include the following:

- Measurement of marketing effectiveness
- Discovery and use (prediction) of specific events as well as the causal relationships among them
- Discovery of emerging trends, needs, and hot spots

Use of user profiles may also raise the accuracy of prediction as well as the probability of discovery.

4.3 Classification of Social Media Applications based on the Purposes

Referring to a book on social data mining [Graubner-Mueller 2011], this section will explain promising business fields based on analysis of social media including the applications described above. As each step of business processes can be considered to correspond to a specific purpose, applications will be enumerated for each purpose. In general, needless to say, business applications are not contrary to individual interests. Rather, such applications are also useful to customers through providing improved services and products in many cases.

In addition, in consideration of usefulness, applications using the generic Web will also be included for reference. Please note that "*" indicates applications mainly using the generic Web and "+" indicates applications using both the generic Web and social media.

(1) Research and development

- *Trend scouting*: A trend survey about a certain known topic which the users frequently describe in social media as well as latent topics with potential values is conducted in order to explore the business environments surrounding development of new products.

- *Consumer behavior analysis*: Investigation of consumer needs, wishes, attitudes, and motives is conducted as to products, product categories, and brands. Since the users' opinions about products, product categories, and brands at large are described in social media irrespective of whether they purchase the products or not, this analysis is useful both for improvement of existing products and development of completely new products which fulfill potential needs.
- *(*) Technology Intelligence*: When a company intends to innovate some product, the company conducts a trend survey on related technologies against specialized technical information repositories on the Web such as patent databases and digital libraries. This is equivalent to investigation of researches by preceding rivals or that of exploratory researches with potential values in research and development of products.

(2) Marketing and sales

- *Product and brand image analysis*: The reputations, sentiments, and opinions about concrete products and brands are analyzed. Indeed, these can be known by conducting a post-purchase survey to users who purchased products. However, since not only the motives and related comments of users who actually purchased the products but also the reasons and opinions of users who thought to purchase the products but did not in reality are described in social media data, analysis of such data is useful for strengthening and changing the current sales strategy.
- *Campaign evaluation*: Some campaigns performed towards users are described by the users in social media data according to the impacts. By analyzing such social data, the effectiveness of marketing can be measured or optimized.
- *Community and opinion leader discovery*: If a community related to a certain product on social media can be discovered, it will be a target of promotion of the product. Moreover, if opinion leaders with large influence in the community can be discovered, it is possible to influence other customers by using channels including the detected opinion leaders in marketing.

(3) Distribution

- *(+) Site and location planning*: Most of information about a certain area as well as customers and rivals within the area has already been published by geographic information services on the Web. On the other hand, reputations about the area or rivals may be described by social media data. By combining such pieces of information, it is possible to accurately choose promising places for opening-a-shop of the company.

(4) Customer services

- *Product recommendation*: Data such as sales histories of a certain commercial product are stored in in-house databases. On the other hand, the ranks and reputations about the product and its relationships with other products are described in social media data. Unifying such data for recommending the product can raise the conversion ratio of the customer as to the product.
- *Customer feedback analysis*: Formal customer feedback is obtained by conducting questionnaires to the users who actually purchased the product. Some dissatisfaction, improvement suggestions, and unexpected uses are described as social data, which can be considered as informal customer feedback. Analyzing such feedback can help to improve the product.

(5) Procurement

- *(*) Content acquisition*: Data in the same categories, such as products, services, and news, spanning over two or more sites are extracted by using Web Service API of each site and aggregated into the unified results.
- *(*) Supplier and price monitoring*: Two or more sites are monitored in an integrated manner so as to compare the suppliers and prices of components for effective supply.

(6) Risk and public relation management

- *Investor sentiment analysis*: It is possible to collect and analyze investors' sentiments as to a specific company through analysis of social media data.
- *Fraud detection*: It is expected that issues that pose a risk to a company such as copyright infringement are uncovered by monitoring file sharing related sites such as BitTorrent sites.
- *(+) Media intelligence*: Collection and analysis of mainstream news of a specific company in generic Web sites and social news and rumors about the company in social media are conducted for customer relationship management. In particular, if bad reputations about the company are discovered in social media in the early stage, they can be used for taking measures so that the situation may not become very serious.

(7) Strategic management

- *(*) Competitive and stakeholder analysis*: For surveillance and analysis of competitors and stakeholders, investigation and analysis are conducted based on data such as management information and news which are published on the official sites and other Web sites.

(8) Human resource management

- *Employer reputation*: The reputation of a company as an employer is investigated based on analysis of social media.
- *(*) Labor market intelligence*: Based on analysis of data on recruitment sites, the labor market relevant to a company is investigated.

4.4 Model Description by the MiPS Model

In this section by concretely considering interactions between social media and the physical real world, the features of a proposed analysis approach to social big data applications will be described, which is one of the concerns of this book.

Events in the physical real world include both artificial occurrences like a producer releasing a new product into a market and natural phenomena like an earthquake occurring somewhere. Moreover, the news about them can also be considered as events in the external world. Thus, such events make up physical real world data in various forms.

On the other hand, individual occurrences such as a customer purchasing a certain product are also a kind of event. However, such events are described only as social data in many cases. Even if the details of the events are stored in some databases (e.g., enterprise databases), it may not be possible to access them from outside of the system. If the users receive any impact from events that happen in the physical real world, they describe comments, reputations, and reactions about the events in social media. Thus they make up social data.

The meta-analysis model called MiPS model includes the following schemas as its basic schemas. Since P and S are schemas, they are denoted by capital letters here (see Fig. 4.3).

In the model of concrete applications, interactions are described by instances of these schemas or combinations of them. Especially interactions between social media and the physical real world are modeled from a viewpoint of influence relationships such as causal relationships, correlations, and spurious correlations. Here some guidance as to how to analyze the interactions and required tasks for such analysis will be explained. The following discussions will start from simple examples of hypotheses and proceed to more complex ones.

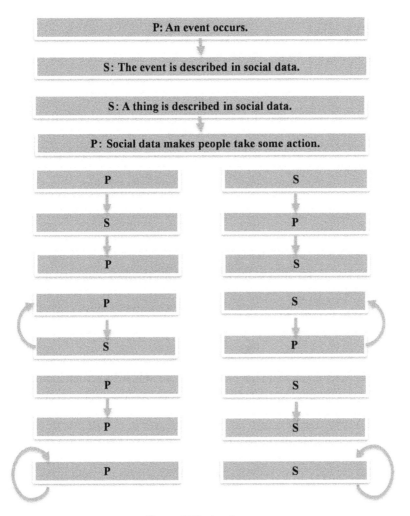

Figure 4.3 Basic schema.

4.4.1 Simple Cases

Firstly, specific hypotheses including social data (S) will be described, focusing on simple cases. The following hypotheses are considered as such examples (see Fig. 4.4).

- (Hs1) After a manufacturer introduces its new product to the market (p), customers who purchase the product describe its reputation in social media (s). p ⇒ s

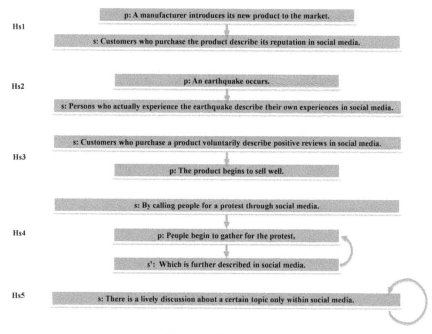

Figure 4.4 Simple cases.

- (Hs2) After an earthquake occurs (p), persons who actually experience the earthquake describe their own experiences in social media (s). p ⇒ s
- (Hs3) After customers who purchase a product voluntarily describe positive reviews in social media (s), the product begins to sell well (p). s ⇒ p
- (Hs4) By calling people to protest through social media (s), people begin to gather for the protest (p), which is further described in social media (s). s ⇒ p ⇒ s * {with loop}
- (Hs5) There is a lively discussion about a certain topic only within social media (s). s ⇒ s * {with loop}

The lower cases p and s represent instances of classes P and S, respectively. Generally, as there are two or more similar instances, they correspond to a set of instances.

Applications which include the following interactions are especially considered to be worthy of analysis from a viewpoint of business applications of social big data.

- a case where business events (physical real world data) trigger posts to social media (social data) like Hs1
- a case where social data trigger physical real world data conversely like Hs3

On the other hand, the case (e.g., Hs4) where events are less related to businesses and the case (e.g., Hs5) where no physical real world data are involved are interesting subjects of analysis from viewpoints of politics or social science. However, such cases will not be discussed any more in this book.

Text mining is one of the tasks required for analysis of hypotheses of our interests such as Hs1, Hs2, and Hs3. Important points in its application to social big data will be explained below.

(1) Topic detection

In the first place, it is necessary to extract information about entities (i.e., topics) such as products (Hs1) and earthquakes (Hs2) from the social contents in order to analyze comments about such entities. In such cases, it becomes important to extract various pieces of information about entities, such as information sources and responses to phenomena, that is, what, when, where, and how users do about entities in addition to information on entities themselves. Especially, as with Twitter, all such related information is not necessarily described as one tweet (i.e., one article). Therefore, it is necessary to identify a coherent set of tweets associated with a certain entity from tweets contained in the user's time line. However, such elements of the set do not necessarily appear continuously on the time line. That is, since various topics cross in the same time line, it is necessary to identify the coherent set, being cautious of the time lag between related articles.

In addition, approaches which focus on the dynamic states of social data are promising for detection of events associated with a topic which is not known beforehand. In relation to events, for example, the number of articles may change temporally or spatially (i.e., geographically) and the relationships between the users may change rapidly. Analysis of such dynamics enables the analyst to know that certain events have happened and to identify topics associated with the events as a result.

(2) Evaluation of importance of articles

In analysis of reputations such as Hs1 or word-of-mouth such as Hs3, it is necessary to apply text mining to the contents about topics described as social data (i.e., articles) in order to evaluate the importance and relevance of the data. Furthermore, in text mining of social data, it is important to use qualitative analysis in addition to quantitative analysis. This is because a sufficient amount of data for quantitative analysis may not be collected depending on topics or applications.

In order to evaluate the importance and relevance of articles as to events, it is effective to focus on the sentiment polarity (i.e., a value ranging from positive to neutral to negative) of terms contained by the contents. That is, the contents are mechanically transformed into continuous variable, ordinal variables (e.g., opinion and reputation), or nominal variables (e.g., product category, event participation, kind of actions) based on the terms contained by the contents if necessary. Especially, values of ordinary variables associated with opinions and sentiments are determined by use of a sentiment polarity dictionary which contains polarities of terms. The quantitative treatment of qualitative variables, such as ordinal variables and nominal variables is generally explained below, referring to quantification theory [Tanaka 1979].

The general way of treating an ordinal variable is described first.

- A binary ordinal variable is regarded as a continuous variable with either 0.0 or 1.0 as its value and is processed as it is.
- For an ordinal variable which has a value in more than two grades (e.g., ++, +, 0, −, − −), the Likert scale is often used. Assuming that there is a latent variable which takes continuous values according to a standard normal distribution, the variable is considered to appear only in five grades.

The conversion method for nominal variables will be explained as follows:

- A binary nominal variable is processed like a binary ordinal variable.
- A multiple-valued nominal variable is processed as follows:

First suppose that the nominal variable A has n categories C_i $(i = 1, n)$. N dummy variables A_i $(i = 1, n)$ are introduced. A_i is a binary variable and has the following values.

If $A = C_i$ then let A_i be 1, otherwise let A_i be 0.

Thus, a nominal variable can be expressed with a group of two or more binary variables.

Furthermore, the following methods can be considered when the social contents contain two or more terms which have sentiment polarity values.

- Evaluation polarities of terms are aggregated by simple statistical methods, such as sum or average.
- As a variation of the above method, evaluation polarities of terms are averaged by using TFIDF in the vector space model as weights.
- In evaluation a set of terms, the relationships between them are taken into consideration. For example, "price is not high", "price is high", and "accuracy is high" are evaluated positive, negative, and positive, respectively.

In this way, as it can be assumed that all variables, whether quantitative or qualitative, take continuous values, it is basically possible to quantitatively analyze the contents.

However, data sufficient for quantitative analysis cannot always be collected. In such a case, it is necessary to qualitatively analyze individual articles, carefully translating the texts into as objective expressions or values as possible. Furthermore, hypotheses for qualitative analysis need to be created depending on applications.

(3) Objective and subjective observations

It is necessary to treat two kinds of observed values about natural phenomena, such as earthquake and weather, in real applications. Thus, observed values include objective values (e.g., seismic intensity and temperature measured by devices) and subjective values (e.g., seismic intensity and temperature felt by humans). In a word, these two kinds of data correspond to physical real world data and social data, respectively. Therefore it is necessary to analyze them by relating to each other. The latter kind of data are often considered more important than the former because they are more local and realistic for the users. Further, they have higher possibility of directly leading to the users' actions.

(4) Description language

In usual applications, it is sufficient to make only a set of social data described in the same language (e.g., Japanese) a target of analysis. However, some applications aim to discover differences by languages. In such cases, of course, it is necessary to analyze social data described in different languages.

For example, according to the research by the author [Ishikawa 2014], places which foreigners frequently visit in Japan are not necessarily in accordance with those which Japanese frequently visit. We collected tweets by specifying major place names in Tokyo spelled by different languages as search conditions. We counted the frequencies of tweets containing each place for each language and ranked places for each language. As a result, we can know the ranks of popular places for people who speak a specific language. Further, by paying attention to user accounts and time, we can discover popular sight seeing routes for foreign visitors, so called golden routes.

4.4.2 More Complex Examples

In this subsection, more complicated hypotheses will be described which combine two simple hypotheses explained above (see Fig. 4.6).

$p_{i,j,k,l}$: **Visit a place**

$s_{i,j,k,l}$: **Post an article**

Task:
Count articles with language, place$_j$, time$_k$, and user$_l$

Figure 4.5 Language, time, and space.

Hc1
p: A manufacturer introduces a new product into market.
S: Responses such as claims about the product are described as social data by the users.
p': The claims make the manufacturer improve the product.

Hc2
p: A manufacturer conducts a campaign of a new product.
s: Users who purchase the product describe their actions and evaluations about the new product in social data.
p': The product starts to sell well due to the articles.

Hc3
p: An academic paper that tomatoes are effective as some measures against metabolic syndromes is published in an academic journal.
s: Persons who get to know about it cite it in social data.
p': Tomatoes begin to sell well and run short in a market due to the social data.

Hc4
p: A research that yogurt has an effect in prevention of influenza becomes some news.
s: The users refer to the news in social data.
p': Yogurt producers increase the amount of production of yogurt according to the social articles.

Hc5
p: Temperature falls suddenly.
s: An increase in the number of users who murmur in social data that "It's chilly."
p': Winter clothing suddenly starts to sell well due to the social articles.

Figure 4.6 More complex cases.

- (Hc1) After a manufacturer introduces a new product into the market (p), responses such as claims about the product are described as social data by the users (s). Furthermore, the claims make the manufacturer improve the product (p′). $p \Rightarrow s \Rightarrow p'$
- (Hc2) After a manufacturer conducts a campaign of a new product (p), users who purchase the product describe their actions and evaluations about the new product in social data (s). Furthermore, the product starts to sell well due to the articles (p′). $p \Rightarrow s \Rightarrow p'$

- (Hc3) After an academic paper that tomatoes are an effective measures against metabolic syndromes is published in an academic journal (p), persons who get to know about it cite it in social data (s). After that, tomatoes begin to sell well and run short in a market (p') under the influence of the paper via the social data. $p \Rightarrow s \Rightarrow p'$
- (Hc4) A research that yogurt has an effect in prevention of influenza becomes some news (p) and the users refer to the news in social data (s). Then yogurt producers increase the amount of production of yogurt (p') according to the social articles. $p \Rightarrow s \Rightarrow p'$
- (Hc5) After the temperature falls suddenly (p) and there is an increase in the number of users who murmur in social data that "It's chilly" (s). Winter clothing starts to suddenly sell well (p') due to the social articles. $p \Rightarrow s \Rightarrow p'$

To analyze these hypotheses requires both mining of natural language texts and discovery of causal relationships at least. The scenario of the analysis will be explained below.

(1) Qualitative analysis

If claims about products are described as social data like the example Hc1, it is important to judge whether the products should be improved based on such claims. In such cases, sentiment polarity analysis of the contents should be carried out as a starting point. However, the frequency of similar claims in social articles which lead to improvement may not always be high. Furthermore, such articles may be outliers of a certain kind. Therefore, analysis only based on the low frequency of similar articles is not sufficient in such cases. Rather, qualitative analysis of such articles becomes more practical than quantitative analysis.

(2) Quantitative prediction and mediation effect

It is necessary to judge the effectiveness of the leverage of social media (s) as to prediction of a phenomenon (p'). In the examples Hc2, Hc3, Hc4, and Hc5, social data (s) correspond to mediation big objects, which are extensions of mediation variables in multivariate analysis. In order to insist that such relationships hold, it is necessary to check how much social data (s) actually raise sales (p'). If any existence of a causal relationship can be confirmed, then the relationship can be used as follows: If seemingly-latent variables (p) can be observed in new articles (s) described by influential users in the same social data as a sensor, then the event (p') which will happen in future may be predicted.

Furthermore, it is sometimes necessary to take into consideration a possibility of an event (p) directly influencing another event (p'). For example, it can be considered as plausible that news of the preventive effect of yogurt over influenza directly causes production increase of yogurt in the example Hc4. In such cases, it is necessary to synthetically consider both the direct effect in the physical real world and the indirect effect in social media.

(3) Sentiment polarity analysis

As already described, in general, it is necessary to perform text mining based on the sentiment polarity values of the contents themselves for analysis of claims or sentiments contained by the contents. Academic journals and news as information sources are discovered by extraction of information from the contents. Especially, it is necessary to analyze subjective observed values (e.g., heat and cold) corresponding to objective observed value (e.g., temperature) for prediction of users' actions.

In addition, in order to find a leading article on a particular topic or its contributor, it is possible to evaluate and use the relevance of the article to the topic, the influence of its contributor to other users and the accuracy of prediction of the article as to the topic.

(4) Open access journal

Recently, open access journals [Laakso et al. 2011] have garnered a lot of attention because they allow everybody to freely access the articles. Usually the period for refereeing is set shorter than that for conventional academic journals. Therefore, scientific discoveries have become known more widely and earlier. As a result, such scientific discoveries are more frequently cited in social data. This fact has partially raised the usability of social data in construction of hypotheses (e.g., Hc3)

In open access journals, not only counts of views and downloads but also citations are available earlier than in traditional journals. The author and his colleagues have performed preliminary experiments so as to make sure that it is possible to discover highly cited papers (i.e., papers with 90 or more citations, HC) only based on similarities as to time series data of views and downloads of papers [Ishikawa 2014]. First we collected 48,261 sample papers with three-month data of downloads from PLoS (Public Library of Science). Next we applied the dynamic time warp CF tree method (an extension of the scalable clustering method BIRCH) in order to cluster the collected data and found a cluster containing a lot of HC papers. In fact, this cluster contains 97.74% of all HC papers (i.e., 389 papers) in the whole set of sample papers. This suggests a possibility that highly cited papers can be discovered with at least 97% confidence by using only 3-month download histories.

Although this is a simple case where a hypothesis consists of p (publish) ⇒ p′ (download) and p as a common cause ⇒ p″ (cite), it is unique in that the case utilizes high speed publishing associated with open access journals (see Fig. 4.7).

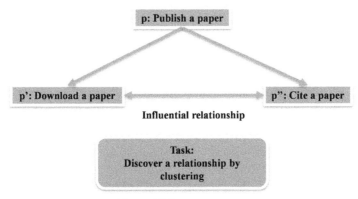

Figure 4.7 Common cause.

4.4.3 Pseudo Correlation Relationships

In this subsection, hypotheses which contain pseudo correlation relationships as influence relationships in addition to causal relationships will be considered. Examples of such hypotheses will be described below (see Fig. 4.8).

- (Hsc1) After an earthquake occurs in the Pacific Ocean (p), customers immediately describe the earthquake (s) in social data and then the tsunami hits the Pacific coast side of Japan later (p′) with high probability. p ⇒ p′ (p ⇒ s)
- (Hsc2) If El Nino occurs (p), it will be a cool summer in Japan (p′) and will be described as social data (s) and next it will become a warm winter in Japan (p″). p ⇒ p′ ⇒ s (p ⇒ p″)
- (Hsc3) If people participate in a popular artist's concert, then most of them take trains and describe their sentiments of the concert in social data. p ⇒ p′ (p ⇒ s)

The approach to processing of these hypotheses will be explained below. Here, positive use of spurious correlations, which should be cautiously treated in discovery of causal relationships in typical multivariate analysis, will be described.

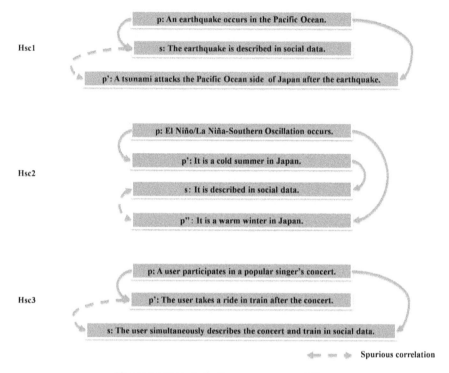

Figure 4.8 Cases including spurious correlations.

(1) Social data as real-time sensors

In the example Hsc1, the phenomenon p is a direct cause of the phenomenon p'. In other words, the phenomenon p becomes a sign of the phenomenon p'. Therefore, such a causal relationship needs to be identified precisely. The phenomenon p' can be predicted if an observed value (p) can be tapped by using social data (s) as a sensor of the value. Since p' is a natural phenomenon, there is of course no strict causal relationship from s to p' although a spurious correlation relationship may hold between s and p'. Furthermore, in this example, the causal relationship from p to p' occurs for an extremely short time. If s can be monitored in near-real time, even if users don't know that the earthquake has really occurred (p), the monitored value can be used as an emergency evacuation alarm from tsunami (p'). Although there is a formal tsunami warning, of course, the warning may not always reach the users before they can escape from the tsunami. Needless to say, as for an earthquake, the extraction of exact information, such as the epicenter, magnitude, and time of occurrence of the earthquake as well as sources of such information, is prerequisite.

(2) Social data as sensor of unknown phenomena

In Hsc2, the phenomenon p is the common cause of the phenomenon p′ (or social data s) and the phenomenon p″. Such causal relationships need to be strictly identified in a similar manner. Although there may be no causal relationship between p′ and p″, pseudo correlation relationships at least hold among them. p″ can be predicted if an observed value (p′) can be tapped by using s as its sensor in this case. In a situation where a phenomenon (p) in particular is not known well yet, it is possible to use such pseudo correlation relationships, not universally but restrictively.

(3) Searching social data for causes of physical real world data

As already described, reasons for events in physical real world data without semantics can be discovered in social data with semantics if such heterogeneous big data sources are appropriately related to each other. Some kinds of optimization in services can be done by using the results as follows (see Fig. 4.9).

In Japan where railroad networks are well-developed, people who go to concerts played by popular musicians (p) mostly ride in trains at the nearest stations when they return home after the concerts end. As a result, the number of passengers as physical real world data (p′) rapidly increases through their IC cards. Moreover, many participants describe their sentiments about concerts as well as crowded stations in social data (s). As a result, such posts increase rapidly in number. On the other hand, those who are in charge of rail transport operations are interested in any cause of such a rapid increase in the number of passengers. In such a case, paying attention to spurious correlations by collecting and analyzing a set of articles as to the same place (e.g., the crowded station) during the same period (e.g., the period of congestion), which abruptly starts to increase in

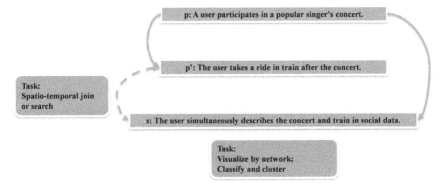

Figure 4.9 Deployment of spurious correlation.

number like a burst, articles posted by concert participants can be found among the set, from which customers' interests (e.g., popular concerts) as a common cause of s and p' can be reached in turn.

Next, by monitoring similar concerts scheduled in home pages of related sites and found as information sources, railway companies can take some measures against train or station congestions in advance. They include some optimization such as effective distribution of passengers to two or more stations or another means of transportation through public campaigns. This can be considered as one of the promising examples which positively use spurious correlations between physical real world data and social data.

Social data in this case contain congestion at a concert site in addition to those at stations. So as to predict future congestions in transportation, it is necessary to precisely extract only transportation-related congestions by applying classification or clustering to social data collected based on spurious correlations [Ishikawa 2014].

Some remarks about the implementation of spurious correlations will be made. Spurious correlations contain temporal and spatial information in many cases. In such cases the spurious correlations could be logically described by specifying conditions on spatial or temporal data as θ-join in SQL. The real implementations, however, require various suitable methods because the units (i.e., granularities) or places (i.e., the contents or tags) of data to be compared may differ, depending on application domains. Generally it is possible to find corresponding parts of heterogeneous non-stream data based on sorting or hashing. In a case of heterogeneous stream data, it is possible to use a method that a subset of one stream data is identified by using information such as data bursts or semantics other than spatial and temporal information and other stream data are filtered by using the spatial or temporal information extracted from the identified subset of the former stream data as conditions.

4.5 Outlook

Future outlook as to what the introduction of the framework for integrated analysis is expected to bring about will be described here. Through the establishment and spread of the introduced framework, some sectors will be able to analyze and deploy big data which other sectors have produced although the same sectors naturally produce and use their own big data. If free markets for big data are formulated and big data in public sectors are widely opened, then the distribution and deployment of big data will be accelerated. Further, big data will intermediate between heterogeneous sectors and will produce a possibility that industrial structures will change dynamically. Eventually, advanced knowledge or wisdom will be constructed throughout our society.

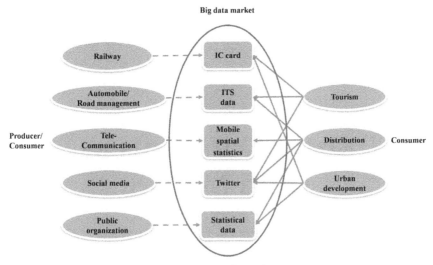

Figure 4.10 Dynamic industrial structure.

References

[Graubner-Mueller 2011] A. Graubner-Mueller: Web Mining in Social Media, Social Media Verlag (2011).

[Ishikawa 2014] H. Ishikawa: social big data science: from transaction mining to interaction mining (invited talk). In Proc. Korea-Japan Database Workshop (2014).

[Laakso et al. 2011] M. Laakso, M. Laakso, P. Welling, H. Bukvova, L. Nyman, B.-C. Björk, T. Hedlund: The Development of Open Access Journal Publishing from 1993 to 2009. PLoS ONE 6(6): e20961. doi:10.1371/journal.pone.0020961 (2011).

[Tanaka 1979] Y. Tanaka: Review of the Methods of Quantification, Environmental Health Perspectives 32: 113–123 (1979).

5

Basic Concepts in Data Mining

This chapter describes basic concepts in data mining, typical tasks for data mining, and basic data structures as targets of data mining.

5.1 What is Data Mining?

First, the fundamental concepts of data mining [Han et al. 2006, Tan et al. 2006], which can be used as principal techniques for constructing hypotheses in the analysis of social big data, will be briefly described. Data mining is, in a nutshell, to discover frequent patterns and meaningful structures appearing in a large amount of data used by applications. Principal techniques, such as multivariate analysis, for validating hypotheses in social big data will be explained in a separate chapter.

One of the basic techniques for data mining is association rule mining, also known as association analysis. It is to discover frequent co-occurrences between structured data used in business applications, which are usually managed by database management systems (DBMS) such as relational database systems. An algorithm called Apriori is used in many cases for that purpose. For example, association rule mining discovers combinations of items co-occurring frequently in a group of items (i.e., contents of the shopping carts) that customers have purchased at the same time in retail stores such as supermarkets. Association rules are made from frequent combinations of items discovered by the algorithm. Based on association rules, a lot of application systems recommend a set of items by revising arrangements of them. Association rule mining is extended and applied to the history of product purchases and the history of click streams on the Web pages in order to discover the frequent patterns of series data. Mining historical data is called historical data mining in particular.

On the other hand, a classifier is learned based on data whose classes (i.e., categories) are known in advance. Then, if there is new data, classes to which they should belong are determined by using the learned classifier. This task called classification is one of the basic data mining techniques. Naïve Bayes and decision trees are used as typical classifiers. Classification is used by such a variety of applications as determination of promising customers, detection of spam e-mails and determination of categories of new specimens in science or medicine. Determination of continuous values such as temperatures and stock prices is also called prediction of future values. Prediction requires methods such as regression analysis as a basic approach or multivariate analysis as a more advanced approach. Indeed, these analytical approaches have been developed more or less independently from data mining. However, they are considered a kind of extensions of data mining and will be described as one of the key technologies for social big data mining separately in this book. Based on a combination of two or more existing classifiers, ensemble learning creates a more accurate classifier than each of the original ones.

It may be possible to define the degrees of similarity between data even if the categories of the data are not known in advance. The opposite concept of similarity is dissimilarity or distance. Based on the defined similarity, grouping data into the same group which are similar to each other in a collection of data is called cluster analysis or clustering, which is also one of the basic technologies of data mining. Unlike classification, clustering doesn't demand that the names and characteristics of clusters are known in advance. Techniques such as a hierarchical agglomerative method and a nonhierarchical k-means method are often used for clustering. Promising applications of clustering include discovery of groups of similar customers for marketing.

A data mining task which can detect exceptional values or values different from standard values is called outlier detection. There are methods for outlier detection based on statistical models, data distances, and data densities. There are alternative ways to find outliers using clustering and classification. Outlier detection has been used for applications, such as detection of credit card frauds or network intrusions.

5.2 Technical Issues and Related Technologies

Here the relationships between data mining and its peripheral technologies will be summarized in order to better understand the features of data mining. Since there are various technologies related to data mining such as databases, information retrieval, and Web search (i.e., search engine), the relationships between data mining and such technologies will be described as follows.

A database is a mechanism for efficiently managing and accessing a large amount of data which social big data applications use. Descriptors for data structures, operations, and constraints dedicated to databases are collectively called a data model. Networks (or graphs) and hierarchical structures (or trees) have often been used as data models since the early days. The former and the latter are called a network data model and a hierarchical data model, respectively.

Nowadays, relations (also called tables), objects, and semi-structured data (e.g., XML) are widely used as data models and are called a relational model, an object-oriented model, and a semi-structured model, respectively.

Furthermore, according to the data models on which databases are based, they are classified into hierarchical databases, network databases, relational databases, object-oriented databases, XML databases, and so on. Generally, a large collection of data as a target of data mining is managed by such databases. The software for managing databases is DBMS.

Data warehouses are similar to databases. Data warehouses unify various sources of information based on concerns such as customers, sales which are needed for decision-making of companies. Transactional databases used by mission-critical tasks are one of the important sources of data warehouses. In order to analyze data generated from such sources of information from a temporal viewpoint, data in data warehouses are usually not updated but just added, therefore, the past data continue to remain as they are. Such data are called time series data. The user makes decisions by mining or analyzing data in data warehouses. Data warehouses are usually built on top of relational databases or dedicated multidimensional databases. In addition, a container which collectively manages related data is sometimes called a repository. For example, pages and links crawled by search engines are stored in dedicated repositories.

Next, in information retrieval, information similar to search terms specified by the user is retrieved based on a vector space model, which consists of feature vectors extracted from text contents. Each component of a feature vector of a document is a value of TFIDF (term frequency × inverse document frequency), which is a feature of each search term contained in a document. Simply put, TFIDF takes into consideration both the frequency of a certain search term within a document and the degree of rareness of the document containing the term in the whole set of documents. On the other hand, a feature vector can also be considered for a search term. In that case, TFIDF of a search term contained in each document is each component of the term vector. The vector space model is based on a matrix called a term document matrix whose columns and rows correspond to these two kinds of vectors, respectively. A query can be expressed by the same feature vector as usual documents by regarding the query as a virtual document which contains only search terms. Ranking two or more documents which satisfy a query can be carried out according to the value of the similarity (e.g., cosine measure using feature vectors) between a virtual document corresponding to the query and all the normal documents.

Generally, based on the analysis of textual contents of documents, classification and clustering of documents can be performed. Such technologies are collectively called text mining or more generally contents mining. On the other hand, analysis of link structures of Web pages is called structure mining or link mining. Analyzing the page contents by contents mining and the link structures characteristic to the Web by structure mining, Web search synthesizes the analysis results and ranks retrieved pages based on them. Especially in application of contents mining to the texts contained in the pages, if a feature term corresponding to a search term appears in the title or anchor text (text on a link) of a page, the value of TFIDF of the term is more highly weighted than usual. In analysis of the link structures of pages, HITS calculates ranks associated with only pages relevant to search results at the time of searches while PageRank calculates the ranks for all the pages on the Web prior to searches. In short, it can be said that Web search is an application of both contents mining and structure mining to the Web.

Compared with traditional technical disciplines such as statistical analysis, machine learning, pattern recognition to which data mining directly owes, the point that especially makes data mining differ from these is inherent awareness of issues about the processing performance of large-scale data. They will be summarized briefly below.

Firstly, global data surrounding us are increasing so rapidly. According to investigation of IDC [IDC 2008], even if several latest years are taken, global data produced and reproduced per year are estimated to have been 161 exabytes in 2006 and to have been around 10 times larger (1.8 zettabytes) in 2011. Data mining cannot but treat such ever-increasing data, that is, big data, from now on.

To the end, algorithms which can work for practical processing time even if data volume increases are needed as well as systems which realize such algorithms. It is generally called linearity as to processing time that the increase in processing time is mostly proportional to the increase in data volume. In other words, linearity ensures that even if data volume increases, the processing time can be maintained within practical limits by a certain means to raise throughput. The capability for an algorithm or its realization system to maintain such linearity is called scalability. Then, how scalability is attained is one of the emergent technical issues in data mining in the era of big data.

Next, with regards to the processing performance of big data, there is another issue called high dimensionality in addition to scalability. In some cases data mining represent target data as objects with many attributes or vectors of many dimensions. For example, the number of the attributes of objects and the dimension of vectors may be very large like those of documents depending on applications. Issues related to such phenomena are sometimes called curse of dimensionality. For example, in collecting the sample data at a fixed ratio at each dimension in such cases, there occurs a problem that the size of the samples increases exponentially with respect to dimensions. Data mining needs to appropriately handle such problems associated with curse of dimensionality.

What data mining considers as emergent issues is not confined to the increase in the size of data or number of dimensions. The complexity of data structures to treat also becomes a problem along with the wide spread of the application fields of big data. Although conventional data mining has mainly targeted structured data, an opportunity to treat graphs or networks (e.g., Web) and semi-structured data (i.e., XML) is increasing along with the development of the Internet. Moreover, the data produced every moment from sensor networks are essentially time series data and positional information is added to time series data if GPS (Global Positioning System) is used. It can be considered that Tweets (i.e., articles in Twitter) are also a kind of time series data. Unstructured multimedia data such as photos, videos, and sounds are also the target of data mining. Furthermore, in a case where the target data of data mining are distributed, issues such as communication costs, data integration, and security are also caused.

5.3 Tasks of Data Mining

As mentioned above, the main tasks of data mining include the following:

- Association rule mining
- Clustering
- Classification and prediction
- Outlier detection

The relations among data mining, database systems, information retrieval, and Web search have already been explained. Here, by explaining points where data mining differs from database systems or information retrieval, the features of data mining will be further clarified.

Both database search and information retrieval allow the user to specify conditions about needed data and to search a set of data for data which satisfy the specified conditions. On the other hand, given a set of data, data mining aims at discovering various structures, relationships, and rules expressing the characteristics of the data. Figuratively speaking, given a set of searched data, data mining tries to discover search conditions for the set. The significant structures, relationships, and rules discovered by data mining are called patterns or models. Based on data (or effects) about observed phenomenon, in general, to explore and discover essences (or causes) of the phenomenon is called an inverse problem in engineering. It may be said that data mining is a kind of inverse problem in that sense.

Data mining can be positioned in the context of broader technologies called knowledge discovery in database (KDD). Generally KDD takes data from data warehouses or databases and creates knowledge as a result. KDD consists of the following steps (see Fig. 5.1).

1. Data cleansing: Remove data (noises) from a data source which disturb consistency.
2. Data integration: Unify two or more data sources if needed and store them in databases.
3. Data reduction and selection: Choose the essential parts of data as a target of data mining from databases. Moreover, reduce the target data to the amount of data which can be practically processed.
4. Data conversion: Change the target data into data structures suitable for data mining.
5. Data mining: Extract patterns from the converted data using intellectual methods as described in this part.
6. Pattern evaluation: Evaluate the extracted patterns according to a certain measure (interest level), and identify as knowledge the patterns which are truly interesting to the user.
7. Knowledge representation: Express and visualize the identified knowledge for its effective presentation.

Thus data mining is one of the essential steps in the process of knowledge discovery. In addition, KDD is not a one-way process but is usually accompanied by feedback loops to any prior step based on the so far acquired knowledge.

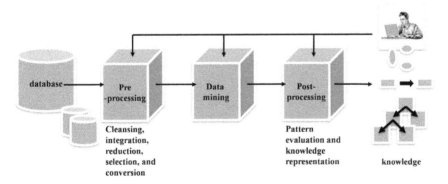

Figure 5.1 KDD processes.

5.4 Fundamental Data Structures

In this section the structures of data which data mining handles as a target will be explained. In the era of big data, the variety of data structures and the velocity of data generation are key issues as well as the size of data. Here, a collection of data is especially called a data set. The fundamental structures of such data will be summarized as follows.

(1) Record

In many cases of data mining, the target data set is represented as a collection of records. The data in such cases are equivalent to records in relational databases (called tuples), that is, structured data. Each record consists of one or more attributes (called column) (see Fig. 2.5). Generally, an attribute or a combination of two or more attributes which can identify a record is called key attributes. In this book, the key that consists only of a single attribute is especially called a record identifier.

First, attributes can be classified into categorical attributes (i.e., qualitative attributes) and numerical attributes (i.e., quantitative attributes).

Categorical attributes can be further classified into attributes whose values have only to be mutually distinguished and attributes whose values have ordinal relationships as well. The former is called nominal attribute, whose values can be compared by an equal mark (=) or non-equal mark (!=). The latter is called ordinal attribute, whose values can be compared as to magnitude in addition to equality.

Generally, according to the kinds of numerical attributes, operations such as addition and multiplication can be performed on values in addition to comparison of them.

Moreover, attributes can be classified also according to the characteristics of the domains of values, especially, the cardinality as a set. An attribute which is expressed with real numbers is called a continuous attribute. On the other hand, if the domain set of attribute values has the same characteristics as a subset of natural numbers, such an attribute is called a discrete attribute. If the number of discrete attribute values is limited, such

ItemID	ItemName	Price\
I_1	German wine	1700
I_2	French wine	2800
I_3	Japanese wine	2100
I_4	Chilean wine	1200
I_5	Italian wine	1500

Figure 5.2 Examples of records.

an attribute is called a multivalued attribute and especially if the value can be expressed by either 0 or 1, the attribute is called a binary attribute or a dichotomous attribute.

Please note that there are multivalued attributes also in the database field, but in such a case the attribute is allowed to have a set (collection) of values as its value. So a multivalued attribute in the database field is called a set attribute or repeating group so as to be distinguished from that in data mining.

For example, both body heat and frequencies are numerical attributes. While body heat is expressed by continuous values, frequencies are expressed by discrete values. On the other hand, a sex such as male and female and a disease situation such as slight and serious are examples of categorical attributes and are generally denoted by discrete values. Furthermore, there exist no order relationships among sexes while there exist order relationships among disease situations. However, such classifications are dependent on application domains.

In addition, throughout this book, the terms variable and attribute are interchangeably used. Relational DBMS are usually used to store collections of records. The definition, querying, and updating of record data can be performed with international standard language called SQL.

(2) Transaction

In particular, if the primary attribute other than the identifier of records contains a set of items, such record data are simply called transactions. Please note that such transactions are not exactly the same as the concept of transactions as a unit of processing in the database field. Combinations of items purchased simultaneously, whether at online stores or real stores are typical examples of transactions (see Fig. 5.3).

TransactionID	*Items*
T_1	I_1, I_2, I_4, I_5
T_2	I_2, I_3, I_5
T_3	I_1, I_2, I_4, I_5
T_4	I_1, I_2, I_3, I_5
T_5	I_1, I_2, I_3, I_4, I_5
T_6	I_2, I_3, I_4

Figure 5.3 Examples of transactions.

(3) Data matrix

If all non-identifier attributes of a record are numeric attributes, record data can be regarded as points on a multidimensional space. In that case, the identifiers of the records correspond to the unique names of the points. If a record excluding its identifier is regarded as a vector, the data set constitutes a data matrix. Please note that if each vector is allowed to correspond to a row (column) of the matrix, the identifier can be substituted by the position of the row (column). Similarly, a data matrix may be made for categorical attributes that have been converted to numeric attributes. In clustering, a distance matrix whose components are distances of data directly calculated from a data matrix are often used.

(4) Sequence data, time series data, spatial data, and spatiotemporal data

Transaction data are explicitly related with time when they are generated. A sequence of two or more transactions is called time series or stream. The sensor data such as infrared rays, temperature, illumination, and carbon dioxide which are obtained through the sensor network laid around a certain place is an example of time series data. Even if there is no explicit information as to time, the order of data may be important. Generally such data are called series or sequence. The base sequence of DNA in bioinformatics is one such example. Data with spatial information such as geographic information instead of time is called spatial data. Furthermore, the spatial data which are generated by moving together with GPS receivers turn into spatiotemporal data. The whole set of data generated by sensor networks may become a kind of spatio temporal data by taking the installation positions of the sensors into consideration even if they are not movable. Tweets containing geographic information and check-in/check-out data in transportation can also be viewed as examples of spatiotemporal data.

(5) Semi-structured data and graphs

In general, data and relationships between them can be denoted by nodes and edges of a graph, respectively. For example, the whole Web or its parts can be modeled with graphs if pages and links between pages are represented as nodes and edges of the graphs, respectively (see Fig. 5.4). Follower-followee relationships and friend relationships in social data can also be represented by graphs. Chemical compounds can be directly expressed by graphs. Furthermore, if the similarity between two pieces of data is transformed into the weight of an edge between nodes corresponding to the pieces, it is possible to model the data set as graphs. Please note that graphs include trees or semi-structured data as special cases. XML, which are used as data formats for describing Web pages or exchanging data on the Internet is an example of semi-structured data type (see Fig. 5.5).

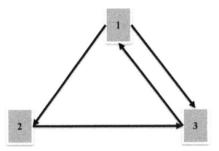

Figure 5.4 An example of a graph.

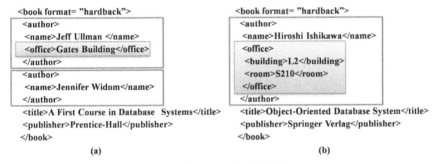

```
<book format= "hardback">
   <author>
      <name>Jeff Ullman </name>
      <office>Gates Building</office>
   </author>
   <author>
      <name>Jennifer Widom</name>
   </author>
   <title>A First Course in Database  Systems</title>
   <publisher>Prentice-Hall</publisher>
</book>
                    (a)
```

```
<book format= "hardback">
   <author>
      <name>Hiroshi Ishikawa</name>
      <office>
         <building>L2</building>
         <room>S210</room>
      </office>
   </author>
   <title>Object-Oriented Database System</title>
   <publisher>Springer Verlag</publisher>
</book>
                    (b)
```

Figure 5.5 An example of XML data.

(6) Media data

Texts range from flat texts with no structures to semi-structured texts formatted by HTML or XML. The features of texts are often vectorized based on the frequencies of characters or feature terms contained by the texts. Moreover, images, videos, and sounds are contained in multimedia data. Multimedia data are described by the primary data (i.e., unstructured raw data) as well as the secondary data (i.e., metadata as annotations to the primary data).

The problems about a data set, which is a collection of data, is described in the end of this section. As mentioned previously, the size (number) of data is greatly related to the scalability of data mining. On the other hand, the number of the attributes of data, also called dimension, will cause the problem of a curse of dimensionality in case of high dimensionality. Both problems are related with the quantity of data.

5.5 Data Quality

In this section, problems related to the quality of data will be discussed.

Big data are inherently vague. Therefore, data mining must treat data that are not always completely correct for various reasons.

First, there may be noises in data sets. The concept which must be distinguished from mere noises are outlying values or outliers. While removal of noises generally leads to better results of data mining, detection of outliers may sometimes lead to discovery of rare and important patterns or phenomenon. However, it is generally difficult to sharply distinguish between both of them. Whether specific data are noises or outliers should be decided depending on application fields.

Moreover, data may include a deficit value. A data set may contain two or more identical data and mutually contradictory data. Since the quality of data may affect the result of data mining largely, such problems must be appropriately treated as follows:

(1) Preprocessing

Such situations about data bring about the necessity of processing data before data mining.

Generally preprocessing data is performed through the following steps.

1. Data cleansing
2. Data integration
3. Data reduction and selection
4. Data transformation

This process will be described step by step.

a) *Data cleansing*
    Imperfections associated with data are resolved by data cleansing. That is, noises are removed by some methods such as prediction, binning, and clustering. In binning, each value is represented with a section where it is contained. When an outlying value is considered to have any bad influence on normal patterns, it is deleted like noises. The attribute or the whole record containing a deficit value is deleted in some cases. In other cases, the deficit values are either complemented by certain values such as common values (i.e., default values) or disregarded. In a case of multiplication of the same data, only one is left and the others are deleted. In a case of inconsistency between data, it is resolved by using knowledge or restrictions of the domain. Instead of removal of noises or outlying values, algorithms for data mining which cannot be easily affected by the existence of these values are invented as alternative solutions.

b) *Data integration*
    Substantially the same data are sometimes distributed and stored in two or more databases. The names, value units, or data structures of the attributes which have the same contents in such cases may differ

from one database to another. Then, such heterogeneities are solved and databases are unified by using metadata (i.e., data about data) of each database.

c) *Data reduction and data selection*
Of course, it is desirable to make the size of big data so small that practical processing can be performed on them. Some of the following methods can be used for that purpose.

- Aggregation (e.g., average and sum) can reduce the data size.
- Sampling can also reduce the data size.
- Discretization of numerical values based on binning or entropies can reduce the number of distinctive values.
- It is possible to delete irrelevant or redundant attributes based on domain knowledge in application fields or analysis by correlations between different attributes.
- A systematic approach can discover optimal combinations of attributes to produce good results by exhaustive search.
- Principal component analysis (PCA) can reduce dimensions (i.e., the number of attributes).

d) *Data transformation*
In order to appropriately combine values of two or more attributes or to emphasize portions of important values, normalization (e.g., min-max, z-score) of values or conversion of values by functions (e.g., absolute value, logarithmic) can be used.

(2) Postprocessing

Postprocessing of data is one of the important processes of KDD like preprocessing of data. While data preprocessing is related to the quality of an input value, data postprocessing is related to the quality of a final result. As data postprocessing, the final result is evaluated in order to find patterns which are truly meaningful in application fields. Then the result is represented as knowledge intelligible to the users. To this end, the result is sometimes visualized.

In the rest of this part, association rule mining, clustering, classification, and prediction will be described as basic technologies for data mining. Then, Web structure-, Web contents-, and Web access log mining will be explained as Web mining, followed by explanation of deep Web mining and information extraction related to it. Further, media mining as to data such as trees, XML, graphs, and multimedia will be described.

References

[Han et al. 2006] J. Han, M. Kamber: Data Mining: Concepts and Techniques, Second Edition, Morgan Kaufmann (2006).

[IDC 2008] IDC: The Diverse and Exploding Digital Universe (white paper) (2008). Available at http://www.emc.com/collateral/analyst-reports/diverse-exploding-digital-universe. pdf accessed 2014

[Tan et al. 2006] P.-N. Tan, M. Steinbach and V. Kumar: Introduction to Data Mining, Addison-Wesley (2005).

6

Association Rule Mining

This chapter explains association rules which can be discovered based on frequent combinations of items. First, the basic concepts and types of association rules as well as the applications will be described. Next, the chapter will explain the basic algorithm for calculating frequent item sets and generating association rules based on them.

6.1 Applications of Association Analysis

First, association rule mining or association analysis as one of the fundamental techniques for constructing hypotheses in big data applications will be explained from a viewpoint of usefulness. Analysis of items that customers buy together in a supermarket is generally called market basket analysis. If a set of items that are frequently bought together can be analyzed, the result can be used for effective exhibition of items.

For example, the following sales strategies can be considered:

- All items contained in the set of frequent items are arranged as near to each other as possible so as to increase an opportunity for the customer to buy them together.
- All items in the set are bundled so as to make the customer buy them together.
- All items in the set are arranged as far from each other as possible so as to increase the opportunities for the customer to buy other items as well as the set, while moving between them.
- One item in the set is sold at a bargain price and the profit ratio of another in the set is raised and then all the items are bundled so as to increase the profit as a total.

Here, a set of items which are contained in a shopping basket (or shopping cart) is called a market basket transaction or a transaction shortly. Please note that items represent not only concrete commodities but also persons, events, pages, terms, and abstract concepts. Caution is also necessary because the concept of transaction here is different from that of transaction in the database field.

In a word, association analysis is to conduct market basket analysis and to discover as the result of analysis, a rule such that if many customers buy one specific item, then they often buy another specific item together. Of course, applications of association rule mining are not limited to market basket analysis. Some applications in search engines discover combinatorial patterns of terms which frequently appear on the same Web page or combinatorial patterns of search terms which are frequently and simultaneously specified in search queries. Other applications discover access patterns (i.e., sequences of visited Web pages) which frequently appear in the access histories of Web servers. Association analysis is also used so as to discover combinations of followees and those of hashtags which are frequently and simultaneously specified in social data. Furthermore, applications of association analysis such as elucidation of interactions in gene expression are spreading quickly in biology.

Therefore, although this chapter explains mining of association rules using daily commodities as items in examples, please note that the technology can be used for a broader range of applications.

In mining association rules, it is necessary to calculate combinations of items, whose number may often be very large. That is because if the number of distinctive items (i.e., elements of a set) is equal to N, the number of combining (subsetting) items will be equal to 2^N. A set which has subsets as its elements is called a powerset of the original set. Thus, the size of the powerset is the order of the exponential function of the size of the original set. So if the original set is denoted by S, its powerset is usually denoted by 2^S. Indeed it may not actually happen that the total number of the combinations becomes so extremely large, considering only the frequent items. In big data applications, however, it is practically very important to design algorithms that can find out such frequent combinations of items efficiently.

6.2 Basic Concepts

First, basic concepts which are necessary for mining association rules will be explained. Here, let I_k be each item and $I = \{I_1, I_2, --, I_n\}$ be a set of all the items. As described above, items may represent not only daily commodities in a shopping basket but also more general concepts such as events, persons, and terms.

Let T be each transaction. Thus, T is a set of items and is contained by $I(T \subseteq I)$. A transaction identifier TID (also called a record identifier, RID) is associated with each transaction. Thus, a transaction is identified by its

TID. Let D be a database, now considered as a target of mining. Then D becomes a set of all transactions.

Using the above basic concepts, an association rule can be expressed as follows:

(Definition) Association rule
- $A \Rightarrow B$ (A implies B)

Here $A, B \in 2^I$ (i.e., A and B are sets of items) and $A \cap B = \emptyset$ (i.e., empty set).

Next, as concepts accompanying association rules, support (more accurately, a degree of support) and confidence (more accurately, a degree of confidence) will be defined. The support s is a ratio of transactions over the database D which include both A and B. Please note that such an item set can be represented as $A \cup B$ but not $A \cap B$. On the other hand, the confidence c is a ratio of transactions containing both A and B over transactions containing A in D. In other words, support is a measure of importance of an association rule while confidence is a measure of reliability of an association rule.

Using a probability P, support and confidence will be put in another way as follows:

(Definition) Support and confidence of an association rule
- Support $(A \Rightarrow B) \equiv P(A \cup B)$
- Confidence $(A \Rightarrow B) \equiv P(B \mid A)$

Here $P(A \cup B)$ and $P(B \mid A)$ represent a probability and a conditional probability, respectively. The latter is the probability that B will happen under the condition that A happened.

Next, strength of an association rule will be defined. When the minimum support denoted by min_sup and the minimum confidence denoted by min_conf are given, an association rule which has support not less than the minimum support and confidence not less than the minimum confidence simultaneously is called a strong association rule.

A set of items is shortly called an itemset. An itemset that consists of k different items is denoted by k-itemset. The frequency of occurrences of an itemset equals the number of transactions which include the itemset. The frequency of occurrences of an itemset is called support count.

If the support count of an itemset is not less than the minimum support $\times |D|$ (called minimum support count), it is said that the itemset satisfies the minimum support. An itemset which satisfies the minimum support is called a frequent itemset or large itemset and it is usually denoted by L_k where k represents the number of items constituting the itemset.

Let the support count of an itemset A be support_count (A). The support and confidence introduced above can be alternatively defined as follows:

(Definition) Support and confidence of an association rule (revisited)

- support $(A \Rightarrow B) \equiv P(A \cup B) = \frac{support\_count(A \cup B)}{|D|}$
- confidence $(A \Rightarrow B) \equiv P(B \mid A) = \frac{support\_count(A \cup B)}{support\_count(A)}$

For example, the association rule that "the customer who buys fish also buys white wine" can be expressed as follows:

- Fish (a customer buys fish) $\Rightarrow$ white wine (the customer buys wine) [support = 50%, confidence = 75%]

Since all the necessary concepts have been defined, how association rule mining is performed will be explained below. That is, mining association rules consists of the following two steps.

1. Frequent itemsets are discovered.
2. Strong rules are generated from the frequent itemsets.

Between the above two steps, the computational complexity of the first step is larger in that the step generally has to handle a powerset of itemsets. Therefore, to efficiently perform the first step is more important in mining association rules.

6.3 Various Association Rules

So far simple association rules for items have been described. There are several kinds of association rules. Generally, association rules can be classified in two or more ways as follows:

(1) Classification by the types of values

An association rule about discrete-valued attributes, which take a limited number of values like the attributes "buy (items)" and "star rating", is called discrete association rule.

The following is an example of a discrete association rule.

- red wine (a customer buys red wine) $\Rightarrow$ cheese (the customer buys cheese)

That is, the attribute "buy" has discrete values such as red wine and cheese.

On the other hand, an association rule about numerical-valued attributes is called numerical association rule. The whole range of numerical values of an attribute is divided into two or more subsections in many cases. Generally, a method to express each value of a numerical attribute by a distinctive section which contains the value is called discretizing. The following is an example of a discretized numerical association rule.

- 25 < age < 30 (a customer is aged 25 < age < 30) ⇒ white wine (the customer buys white wine)

Here the left side of the above rule expresses a section (more exactly, its identifier, i.e., discrete value) as a value of the attribute age.

(2) Classification by dimensions

An attribute which appears in an association rule is considered to represent a dimension. A rule can be classified according to dimensionality. The following is an example of a one-dimensional rule which consists only of a single attribute "buy".

- red wine (a customer buys red wine) ⇒ cheese (the customer buys cheese)

On the other hand, the following is an example of a multi-dimensional rule which consists of two attributes "buy" and "age".

- 25 < age < 30 (a customer is aged 25 < age < 30) ⇒ white wine (the customer buys white wine)

(3) Classification by degrees of abstraction

A collection of two or more rules, called a rule set, will be considered. If a hierarchical relationship based on degrees of abstraction can be considered between items contained by a rule, such a rule set is called a multi-level rule set. For example, the following rule set is multilevel.

- 20 < age < 45 (a customer is aged 20 < age < 45) ⇒ wine (the customer buys wine)
- 30 < age < 45 (a customer is aged 30 < age < 45) ⇒ red wine (the customer buys red wine)

Conceptually, wine is more general than red wine and red wine is more specific than wine conversely. On the other hand, a rule set where there is no such hierarchical relationship between items contained by rules is called a single-level rule set. Such hierarchical relationships are usually managed by a dedicated knowledge base.

(4) Classification by data structures

Discovering frequent itemsets is classified into two separate tasks according to whether an order of transactions is considered or not. The former is mining association rules for series data. The latter is mining association rules for normal unordered data. Furthermore, mining of association rules may be extended to that of more complex data structures such as trees and graphs.

6.4 Outline of the Apriori Algorithm

Here the algorithm called Apriori [Agrawal et al. 1993] will be introduced as a fundamental algorithm which efficiently mines association rules. This algorithm aims at discovering frequent itemsets efficiently. In other words, it tries to evade checking spurious frequent itemsets.

Here the basic algorithm which discovers unordered frequent itemsets and generates discrete, single-dimension, and single-level association rules will be described for the sake of simplicity. All we have to do in order to generate the other types of association rules is to consider extensions to the basic algorithm.

The following principle holds for frequent itemsets.

(Principle) Apriori
• All the subsets of frequent itemsets are frequent.

More generally, the following monotonically decreasing principle, called downward monotonicity, holds for support counts.

(Principle) downward monotonicity
• $X, Y \in 2^I$ and $X \subseteq Y \Rightarrow$ support_count $(X) >=$ support_count (Y).

One of the salient features of the Apriori algorithm is that it repeats the procedure of looking for $(k + 1)$ itemset using k-itemset, based on the above-mentioned principle (Apriori). That is, first it finds the frequent 1-itemset L_1. Next it finds the frequent 2-itemset L_2 using L_1. Furthermore, it finds the frequent 3-itemset L_3 using L_2. This procedure is repeated until the frequent k-itemset L_k is no longer found. All frequent itemsets $L = \cup_k L_k$ can be found as a result of the algorithm.

The Apriori algorithm fundamentally repeats the procedure which consists of the following two steps.

(1) Joining step

If the frequent $(k - 1)$-itemset L_{k-1} is obtained, the algorithm can find the candidate itemset C_k using the frequent k-itemset L_k. Here let l_1 and l_2 be the elements of L_{k-1}.

Furthermore, let $l_i [j]$ denote the j-th item of l_i. In addition, assume that within an itemset and a transaction, items are sorted in a lexical order (i.e., order of entries of a dictionary). Here let L_{k-1} be a database which consists of $(k - 1)$ fields.

Then, the join (L_{k-1}, L_{k-1}) can be considered to be a natural join operation of the same table by the first $(k - 2)$ fields as the join keys.

The join predicate in this case can be specified as follows:

• $(l_1 [1] = l_2 [1])$ AND $(l_1 [2] = l_2 [2])$ AND ... AND $(l_1 [k - 2] = l_2 [k - 2])$ AND $(l_1 [k - 1] < l_2 [k - 1])$

Here the last condition ($l_1[k-1] < l_2[k-1]$) in the above predicate specifies that the candidate itemsets l_1 and l_2 are different.

A k-itemset $l_1[1] l_1[2]...l_1[k-2] l_1[k-1] l_2[k-1]$ is obtained as a result of the join operation. The above operation will be the self-join of the same table L_{k-1}, which is expressed by the following SQL command.

- INSERT INTO C_k
  SELECT $p.item_1, p.item_2, ... ,p.item_{k-1}, q.item_{k-1}$
  FROM L_{k-1} p, L_{k-1} q
  WHERE $p.item_1 = q.item_1$ AND $p.item_2 = q.item_2$ AND ... AND $p.item_{k-2} = q.item_{k-2}$ AND $p.item_{k-1} < q.item_{k-1}$;

(2) Pruning step

The candidate set C_k includes all the frequent k-itemsets L_k. In other words, C_k also may include itemsets which are not frequent. Therefore, it is necessary to eliminate (i.e., prune) all the infrequent itemsets so as to calculate only L_k. For that purpose, first, it is necessary to calculate the frequency of C_k (i.e., support count) in D by scanning the database D. Furthermore, it is necessary to confirm that the candidate set satisfies the minimum support.

Here it will be shown that the Apriori principle can help reduce the size (the number of elements) of C_k. That is, since the principle assures that the itemset included in C_k is not frequent if at least one $(k-1)$-itemset as a subset of the itemset is not contained in L_{k-1}, it can be deleted from C_k.

Then, the outline of the Apriori algorithm is shown by pseudo-codes below.

(Algorithm) Apriori

```
1.     L₁ ← frequent 1-itemset;
2.     k ← 2;
3.     while (NOT Lₖ₋₁ = Ø) {
4.            Cₖ ← Generate-Candidates (Lₖ₋₁, min_sup);
5.            for (all transaction t ∈ D){
6.                   Cₜ ← Cₖ part of t;
7.                   for (all c ∈ Cₜ){
8.                          c.count ← c.count+1;
9.                   }
10.           }
11.           Lₖ ← {c ∈ Cₖ | c.count >= min_sup};
12.           k ← k + 1;
13.    }
14.    Return ∪ₖLₖ;
```

(Algorithm) Generate-Candidates (L_{k-1}, *min_sup*)

1.　　for (all $l_1 \in L_{k-1}$){
2.　　　for (all $l_2 \in L_{k-1}$){
3.　　　　if $((l_1[1] = l_2[1])$ AND $(l_1[2] = l_2[2])$ AND…AND $(l_1[k-2] = l_2[k-2])$ AND $(l_1[k-1] < l_2[k-1]))$ {
4.　　　　　$c \leftarrow join(l_1, l_2)$;
5.　　　　　if (any $(k-1)$-itemset as a subset of c is not frequent) {
6.　　　　　　Delete c;
7.　　　　　}else{
8.　　　　　　Add c to C_k;
9.　　　　　}
10.　　　　}
11.　　　}
12.　　}
13.　　Return C_k;

Now each step of the Apriori algorithm will be explained. Step 1 of this algorithm finds a frequent 1-itemset. In the repetition after Step 3, Step 4 generates the candidate set C_k based on L_{k-1} first. At this time, pruning of C_k is carried out based on the Apriori principle beforehand. Next the frequency of C_k is counted up for all the transactions at Step 5 to 10. Finally at Step 11, if C_k satisfies the minimum support using the support count, such C_k that satisfy the minimum support is added to L_k.

In order for the algorithm to terminate, the repetition needs to be completed someday.

The terminal condition of the repetition is that L_k become an empty set, i.e., no more frequent itemset be found. Since the database, of course, has a limited number of data and L_k will definitely become empty someday, the algorithm definitely terminates. If the repetition terminates, a union of L_k will finally be calculated for all k at Step 14 to return the set L.

In the algorithm (Generate-Candidates) which generates the candidate set C_k, after the join possibility of two l_{k-1} is checked at Step 3, those two are actually joined. Then, whether all the $(k-1)$–itemset as subsets of the joined result are contained in L_{k-1} is checked at Step 5, and only if all of them are contained, the result can be added to C_k based on the Apriori principle.

Using a fictitious example of a transaction database (see Fig. 6.1), how the Apriori algorithm actually works is illustrated (see Fig. 6.2). Each transaction contains wines (precisely, the identifiers) as a set as shown in Fig. 6.1. The result of each scan of the database is represented by a pair of an itemset and its support count as shown in Fig. 6.2.

ItemID	ItemName
I_1	German wine
I_2	French wine
I_3	Japanese wine
I_4	Chilean wine
I_5	Italian wine

TransactionID	Items
T_1	I_1, I_2, I_4, I_5
T_2	I_2, I_3, I_5
T_3	I_1, I_2, I_4, I_5
T_4	I_1, I_2, I_3, I_5
T_5	I_1, I_2, I_3, I_4, I_5
T_6	I_2, I_3, I_4

Figure 6.1 Transaction database.

First scan	
I_1	4
I_2	6
I_3	4
I_4	4
I_5	5

Second scan	
I_1, I_2	4
I_1, I_3	2
I_1, I_4	3
I_1, I_5	4
I_2, I_3	4
I_2, I_4	4
I_2, I_5	5
I_3, I_4	2
I_3, I_5	3
I_4, I_5	3

Third scan	
I_1, I_2, I_4	3
I_1, I_2, I_5	4
I_1, I_4, I_5	3
I_2, I_3, I_5	3
I_2, I_4, I_5	3

Fourth scan	
I_1, I_2, I_4, I_5	3

Figure 6.2 Execution of Apriori.

Let the minimum support count be 3 (i.e., the minimum support = 50%). C_2 minus the infrequent itemsets (i.e., underlined itemsets) become L_2 with the second scan. All the frequent itemsets are obtained with four scans in a similar way.

As shown in Fig. 6.3, all of the itemsets constitute a lattice, based on set-inclusive relationships between them. In Fig. 6.3, if two itemsets depicted by elliptic nodes are connected by an edge, the upper itemset is set-included by the lower itemset. That is, they correspond to a subset (i.e., upper one) and a superset (i.e., lower one), respectively. For the sake of simplicity, for example, I_1 is denoted just as 1. This lattice lets us know promptly whether an itemset is frequent or not. For example, if $\{I_1, I_3\}$ are not frequent, then

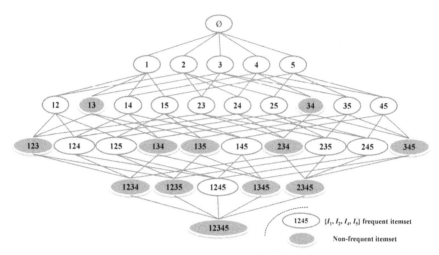

Figure 6.3 A lattice of itemsets.

all the superset such as $\{I_1, I_2, I_3\}$ is not frequent. Conversely, if $\{I_1, I_2, I_4\}$ are frequent, all the subset such as $\{I_1, I_2\}$ is frequent.

The efficiency of such checks of subsets can be increased by managing candidate sets of frequent itemset by a hash tree (see Fig. 6.4). That is, an itemset is stored in the leaf node of the hash tree and a hash table is stored in the nonleaf node. Each bucket of the hash table contains a pointer to the child node. For the sake of simplicity, the following hash function will be considered.

- $H(item) = \text{mod} (\text{ord}(item), 3)$

Here $\text{ord}(i_k) = k$ and mod (m, n) = remainder of the division of m by n.

That is, the downward progress to a child node from a node at depth d from the root in the hash tree is determined according to the result of applying the hash function to the d-th element of the itemset. Please note that d begins with 0. If itemsets exist in the leaf node which searching has reached, they are compared with the current itemset. If they match with

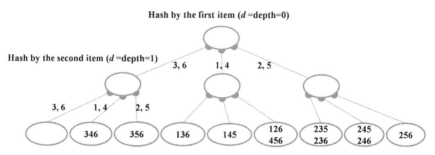

Figure 6.4 A hash tree of 3-itemsets.

each other, the frequency is counted up. Otherwise, the itemset is newly inserted to the leaf node and its frequency count is initialized to 1. Thus, it is possible to evade spurious comparison with candidate frequent itemsets.

6.5 Generation of Association Rules

Once frequent itemsets are discovered by the Apriori algorithm described above, generation of association rules using them is rather straightforward. In order to obtain a strong rule, it is necessary just to calculate the confidence of the rule by the previously defined confidence formula. Association rules are obtained based on the result as follows.

1. Let a proper subset of each frequent k-itemset l except an empty set be s. The number of s is 2^{k-2}.
2. $s \Rightarrow (l - s)$ is created and the confidence is calculated by the following confidence formula.
   - confidence $(A \Rightarrow B) \equiv P\,(B\,|\,A) = \frac{support\_count(A \cup B)}{support\_count(A)}$

That is, in the above formula, let A and B be s and $l - s$, respectively.

If support_count(l)/support_count(s) is larger than or equal to the specified minimum confidence (min_conf), this rule will be a final rule.

Since the rules obtained in this way are made from the frequent itemsets, which means that they satisfy both the minimum support and the minimum confidence, they are strong association rules.

For example, the following six rules are generated from the frequent itemset $\{I_2, I_3, I_5\}$.

- $I_2 \cup I_3 \Rightarrow I_5$ (3/4 = 75%)
- $I_2 \cup I_5 \Rightarrow I_3$ (3/5 = 60%)
- $I_3 \cup I_5 \Rightarrow I_2$ (3/3 = 100%)
- $I_2 \Rightarrow I_3 \cup I_5$ (3/6 = 50%)
- $I_3 \Rightarrow I_2 \cup I_5$ (3/4 = 75%)
- $I_5 \Rightarrow I_2 \cup I_3$ (3/5 = 60%)

Only rules whose confidence is higher than or equal to the minimum confidence (for example, 70%) turn into strong rules.

The computational cost of the Apriori algorithm will be considered here. Let the size of the longest frequent itemset be MAX_k, then the computational complexity required to discover all association rules is $O(MAX_k \times |D|)$, substantially equal to the cost of database scans. The I/O (i.e., input/output) cost between in-memory pages and secondary memories, which determines the actual processing time, can be obtained by dividing this value by the blocking factor (i.e., the number of records per page).

There exist a lot of works which have improved the Apriori algorithm or extended it for better performance [Tan et al. 2006]. In order to reduce the number of database scans, it is possible to apply efficient ways of counting support counts such as DIC (Dynamic Itemset Counting) or efficient data structures such as DHP (Dynamic Hashing and Pruning) and vertical formats, or to make a data set so small using random sampling or partitioning that the whole database may be stored in the main memory. Moreover, methods such as the FP-Growth algorithm to find frequent itemsets without using the Apriori algorithm have also been developed. Furthermore, basic rules have been extended to rules of numerical data and series data. Approaches to increasing the scalability of the Apriori algorithm will be explained together with those to increasing the scalability of other data mining techniques in a separate chapter.

References

[Agrawal et al. 1993] Rakesh Agrawal, Tomasz Imielinski and Arun Swami: Mining association rules between sets of items in large databases. In Proc. of ACM SIGMOD Intl. Conf. on Management of data, pp. 207–216 (1993).
[Tan et al. 2006] P.-N. Tan, M. Steinbach and V. Kumar: Introduction to Data Mining, Addison-Wesley (2005).

7

Clustering

This chapter explains the applications, data structures, and concepts of distances as to clustering for grouping similar data. Then the chapter describes the basic algorithms for producing clusters as well as the methods for evaluating the results.

7.1 Applications

First, applications of clustering will be explained before explaining clustering [Han et al. 2006, Tan et al. 2005] itself. Clustering can be applied to a wide variety of social big data applications as a fundamental technique for constructing hypotheses like association analysis. Let us consider, for example, that customers are grouped based on the similarity of the purchase histories of the customers. If any commodity is purchased often by the customers in the same group, then the commodity can be recommended to customers in the group who have not yet purchased it. Grouping data based on a certain similarity in this way is called clustering. In a sense, clustering is similar to classification in that both group data. While a class (i.e., category) to which data should belong is known in advance in classification, there is usually no assumption about such a class in clustering. Therefore, grouping in clustering, in that sense, should be called partitioning rather than classification. Mathematically speaking, a partition of a set is a subset with no elements in common with one another in the set. Therefore, the set union of all the partitions is equal to the original set. In particular, groups which are created in clustering are called clusters. The names or characteristics of individual clusters are assumed to be unknown in advance.

In addition to business applications such as described above, clustering is used in various applications, such as grouping of social data prior to detailed analysis, grouping of Web search results and Web access histories, the discovery of genes with similar functions, sub-classification of species in biology and patients in medicine. The basic data structures and algorithms used in clustering will be explained in the following sections.

7.2 Data Structures

Before explaining clustering itself, the fundamental data structures for clustering will be explained as preparation. Data structures suitable for describing the target objects of clustering are required. The following data structures can be considered for this purpose.

- Data matrix
- Dissimilarity matrix

The data matrix expresses objects themselves while the dissimilarity matrix expresses differences (i.e., dissimilarities) between objects. Usually, the degree of dissimilarity can be computed from the data matrix by a certain means. Therefore the dissimilarity matrix is often used, which is explained here.

- $[d_{ij}]$

d_{ij} represents the degree of dissimilarity between the two objects i and j, or the distance between them in its general meaning. Therefore, a dissimilarity matrix is also called a distance matrix. If two objects are different, then the degree of dissimilarity is large, otherwise small.

The distance d_{ij} is especially called distance function if it satisfies the following characteristics (i.e., axioms of distances):

- $d_{ij} >= 0$ (non-negativity)
- $d_{ij} = d_{ji}$ (symmetry)
- $d_{ii} = 0$ (identity)
- $d_{ij} <= d_{ik} + d_{kj}$ (triangle inequality)

In other words, some distances satisfy the above-mentioned axioms and others don't. Anyhow most of the clustering algorithms are based on distances. The concrete methods of computing distances will be described later. Here, the relation between similarity and dissimilarity is explained. The degree of dissimilarity can be translated into that of similarity by some methods such as monotonically decreasing linear functions. For example, let the degree of similarity be s_{ij}. Then the degree of similarity s_{ij} can be expressed by the degree of dissimilarity d_{ij} as follows:

- $s_{ij} = 1 - d_{ij}$

In this book, the degree of similarity and the degree of dissimilarity (or distance) are used in the above sense and among them, the more suitable one will be chosen depending on the local contents in the following explanations.

7.3 Distances

Here, a concrete distance which becomes the foundation of dissimilarity will be explained. First the distance of the object which consists of numerical attributes (i.e., numeric variables) is considered. For example, height and weight are both numerical attributes. For simplicity, it is assumed that numerical attributes in consideration can be measured with linear methods. When an object consists of two or more features, it can be expressed as a feature vector.

Typical distances between two feature vectors include the Euclidean distance, Manhattan distance, and Minkowski distance.

The Euclidean distance is calculated by the following formula.

- $d_{ij} = \sqrt{\sum_{k=1}^{m} \left| x_{ik} - x_{jk} \right|^2}$

The Manhattan distance corresponds to the usual distance it takes us to move by car or walk in grids of streets and is calculated by the following formula.

- $d_{ij} = \sum_{k=1}^{m} \left| x_{ik} - x_{jk} \right|$

The Minkowski distance is given by the following formula assuming that p >=1 and it is a generalization of the above two distances.

- $d_{ij} = \sqrt[p]{\sum_{k=1}^{m} \left| x_{ik} - x_{jk} \right|^p}$

The Minkowski distance, especially, becomes the following as p approaches infinity.

- $d_{ij} = \max_{k=1,m} \left| x_{ik} - x_{jk} \right|$

Let v be a vector. The Minkowski distance between v and the zero vector (or the distance between the starting point and ending point of the vector) is called p-norm of the vector and is denoted as follows.

- $\|v\|_p$

In short, it is the length of the vector. $\|v\|_1$ and $\|v\|_2$ are special cases of the p-norm. These are distance functions because they satisfy the axioms of distances.

Next other distances will be explained.

(1) Dichotomous variable

Let's consider a case where each object is denoted by two or more dichotomous variables (i.e., variable whose value is either 0 or 1). The distance of two such objects is defined.

Let two objects be o_1 and o_2. n_{ij} is defined as follows:

- n_{11}: the number of corresponding variables which are 1 both in o_1 and o_2
- n_{10}: the number of corresponding variables which are 1 only in o_1
- n_{01}: the number of corresponding variables which are 1 only in o_2
- n_{00}: the number of corresponding variables which are 0 both in o_1 and o_2.

Then, the distance between dichotomous variables can be defined as follows:

- $$d_{ij} = \frac{n_{10} + n_{01}}{n_{11} + n_{10} + n_{01} + n_{00}}$$

Here, if the value of a variable being 0 means that it is less important than its being 1, this distance is degenerated into the following.

- $$d_{ij} = \frac{n_{10} + n_{01}}{n_{11} + n_{10} + n_{01}}$$

Such a case is called asymmetric while a case where 0 and 1 are equivalent is called symmetrical. A contingency table about two objects is often used so as to calculate the distance of dichotomy variables (see Fig. 7.1).

	$O_1: 0$	$O_1: 1$
$O_2: 0$	n_{00}	n_{10}
$O_2: 1$	n_{01}	n_{11}

Figure 7.1 Contingency table.

(2) Multi-valued variable

If a variable can take three or more values, such a variable is generally called a multi-valued variable. Let N be the total number of multi-valued variables and let T be the number of corresponding variables which have the same values, the distance between multi-valued variables is expressed as follows:

- $$d_{ij} = \frac{N - T}{N}$$

(3) Ordinal variable

Let r be a rank and let M be the total number of distinctive ranks. If the order (i.e., rank) is especially important for multi-valued variables, the value of the rank is normalized by M as follows:

- $\dfrac{r-1}{M-1}$

Based on such normalized values, the distance is calculated as a Euclidean distance introduced for the numeric variables.

(4) Nonlinear variable

A nonlinear-valued variable can be treated in the same way as an ordinal variable after it is converted into a linear value by a certain function. For example, if a nonlinear-valued variable can be approximated by an exponential function, then it can be converted into an ordinal variable by using a logarithm function.

Of course, types of available conversion methods depend on application domains. It is important to choose appropriate distances according to data or applications.

Distances other than the above described distances will be introduced in places where they are necessary.

7.4 Clustering Algorithms

Here, the definition of clustering and the kinds of clustering algorithms will be explained. First the definition of clustering will be described. Assume that the database D contains n objects. Clustering divides D into the $k(<= n)$ groups called clusters c_i. The created clusters must satisfy the following conditions.

- Cluster conditions
  1. Each object belongs to a certain cluster.
  2. One object does not belong to two or more clusters.
  3. There is no cluster which contains no objects.

Notes should be made about the condition (2). Clustering where one object belongs to exactly one cluster is called hard clustering, exclusive clustering, or non-overlapping clustering.

On the other hand, clustering where one object may belong to two or more clusters is called soft clustering, non-exclusive clustering, or overlapping clustering. Especially, if one object belongs to all the clusters with weights (e.g., values from 0 to 1) like fuzzy sets, such clustering is called fuzzy clustering.

In hard clustering, the cluster c_i ($i = 1, --, k$) becomes a partition of D. Such a situation can be described as follows:

- $D=c_1 \cup c_2 \cup \ldots \cup c_k, c_i \cap c_j = \emptyset \ (i! = j)$

Just as clustering has been defined, clustering algorithms will be described. The algorithms for clustering can be divided roughly into two types as follows:

- Partition-based clustering
- Hierarchical clustering

These two types of algorithms will be explained below.

7.5 Partition-based Clustering

The k-means method is one of the typical partition-based methods, which will be explained in detail in the rest of this section.

In the k-means method, the similarity of clusters is measured based on the average (i.e., means) or the centroid (i.e., center of gravity) of values of objects in clusters. Clustering is done so that the similarity of objects within a cluster is larger than the similarity of objects across different clusters.

The algorithm can be described as follows.
(Algorithm) The k-means method

1. Choose k arbitrary objects and let them be the initial centroids of k clusters;
2. Repeat the following procedures until a collection of clusters made by rearrangement of objects remains the same as the previous collection of clusters {
3.   Assign each object to a cluster whose centroid is the nearest to it;
4.   Recalculate the centroid of each cluster so as to reflect newly assigned objects }

Here let us define a squared error as follows.

- $$\sum_{i=1}^{k} \sum_{p \in c_i} |p - m_i|^2$$

Here m_i denotes the centroid (average) of the cluster c_i.

The k-means method creates clusters so as to make this value (locally) minimum. The terminating condition ("there is no change in a collection of clusters") of Step 2 may be replaced with the condition that the decrease in the value of the squared error is smaller than the specified threshold.

How the k-means method works will be explained through a concrete example. Red wines made from Pino noir grapes look bright purple, emit a gay scent, and taste softly acidic. These kinds of wines are produced at various places all over the world. Let us consider clustering wines o_i ($i = 1, --, 5$) as follows:

- o_1: Germany
- o_2: Burgundy (France)
- o_3: Oregon (U.S.A.)
- o_4: Australia
- o_5: California (U.S.A.)

Here, the mutual distances have been subjectively calculated based on the tastes of typical wines of each region though these are fictitious data. Let o_4 and o_5 be the initial centroid with k (= 2). Figure 7.2 illustrates the generation process of clusters by the k-means method.

Consider clustering n data by the k-means method. Let r be the times of repetitions. Then, on the whole, the computational cost will be O (nkr). Note that r and n are small numbers which do not depend on k so greatly. The I/O cost of database access can be obtained by dividing this by the blocking factor.

The k-means method is simple and clear especially in that a center of cluster corresponds to a geometrical concept, that is, a centroid. However, at least the following problems remain to be solved.

- The result of clustering depends on initial arrangements of centroids of clusters.
- The resultant set of clusters is one of local optimal solutions.

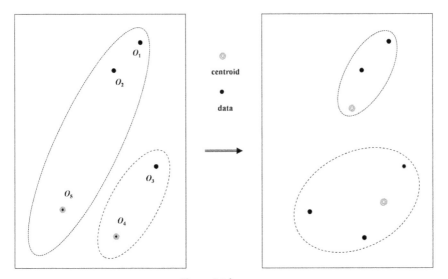

Figure 7.2 k-means.

- The suitable number of k is unknown beforehand.
- The result is sensitive to outlying values.
- The method can be applied only to numerical attributes.
- There is no guarantee that the resultant clusters are balanced.

Some of the above issues have been provided with solutions. The k-medoids method, which is developed as an improvement to the k-means method can handle non-numerical attributes. Instead of the centroids in the k-means method, the k-modoids uses representative objects nearest to the centroids in cluster assignment although the algorithms are basically similar to each other. Moreover, the users don't need to provide a specific value as x in the x-means method in advance. The x-means method creates an appropriate number of clusters by repeating the k-means method so as to optimize a certain evaluation index.

Once clusters are made, next thing to do is to represent individual clusters. Although it is sufficient for some applications just to display objects belonging to clusters, it is generally desirable to aggregate objects in each cluster in order for the analyst to better understand the clusters. The methods for aggregation are as follows:

- Use the centroid of the cluster.
- Use frequent data within the cluster.
- Build a classifier (e.g., decision tree) which can distinguish data within the cluster from data outside the cluster and interpret the classifier.

7.6 Hierarchical Clustering

Hierarchical clustering is the method for making clusters of tree structures. On the other hand, the partition-based clustering methods build flat clusters with no hierarchy between clusters. Of course, the clusters made by hierarchical clustering must satisfy the above described clustering conditions similar to those made by partition-based clustering.

Furthermore, methods for hierarchical clustering can be classified into the method of making a tree of clusters from the leaves toward the root along with the tree structure and the method of making a tree of clusters from the root toward the leaves. The former is, so to speak, a bottom up approach, which continues to merge similar clusters starting with a state where each cluster contains only one object until the total number of clusters reaches the desirable number. Thus, the number of clusters decreases gradually. This approach is called Hierarchical Agglomerative Clustering (HAC).

The latter, on the other hand, is a top down approach, which repeats division of clusters starting with a state where only one cluster contains all data until the total number reaches the desirable number. Therefore, the total number increases gradually. This method is called Hierarchical Divisive Clustering (HDC).

These two have completely reverse directions of the growth processes of clusters. Both the approaches will be explained below.

(1) HAC

HAC has the following algorithm. Let n be the total number of objects and let k ($<= n$) be the desired number of clusters.

(Algorithm) HAC

1. Make the set C of clusters each of which contains one object;/* At this time $|C| == n$;
2. Repeat the following procedure until $|C| == k${
3. Choose the two most similar clusters c_1 and c_2 out of C and delete them from C; /* The similarity measure will be described later;
4. Make a new cluster c_3 which consists of $c_1 \cup c_2$ (i.e., merge c_1 and c_2 into the cluster c_3) and add it to C;}

The tree structure built as a result of HAC is called a dendrogram. If clusters are arranged along the horizontal axis and similarity (or distance) is assigned to the vertical axis, merging clusters begins with two clusters with the largest similarity. In other words, the later the clusters are merged, the less similar they are. Figure 7.3 shows an example of a dendrogram.

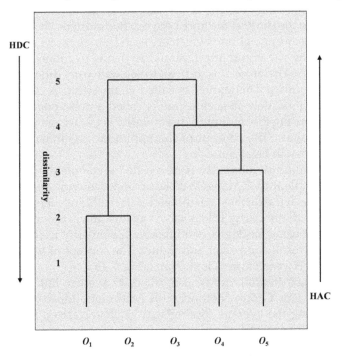

Figure 7.3 Hierarchical clustering and dendrogram.

In the above described algorithm, the terminating condition ($|C| == k$) may be changed into the following condition.

- The condition that there is only one cluster ($|C| == 1$).

All the intermediate clusters and similarities are memorized. Then a set of clusters is determined as a final result just before the currently largest similarity is less than a given threshold.

Next the similarity of clusters will be described. Thus the measure as to the similarity (or the dissimilarity) of two clusters will be considered. If clusters are similar, the degree of similarity is large or the degree of dissimilarity is small. Here the degree of dissimilarity will be focused on.

Let the number of objects contained by each of the three clusters c_1, c_2, and c_3 be n_1, n_2, and n_3, respectively. Moreover, let the dissimilarity between two objects p_1 and p_2 be d (p_1, p_2). The dissimilarity between the clusters c_1 and c_2 can be measured using either of the following criteria.

- The minimum dissimilarity: $min_{p_1 \in c_1 \text{ and } p_2 \in c_2}\{d \ (p_1, p_2)\}$
- The maximum dissimilarity: $max_{p_1 \in c_1 \text{ and } p_2 \in c_2}\{d \ (p_1, p_2)\}$

- The average dissimilarity: $\dfrac{1}{n_1 n_2}\sum_{p_1 \in c_1, p_2 \in c_2} d(p_1, p_2)$

The minimum dissimilarity is also called a single link (SLINK), which is equivalent to the shortest distance between two clusters, i.e., the distance between the nearest neighbors which are each from separate clusters. The clustering using the minimum dissimilarity tends to connect a series of interim clusters. Therefore, it is difficult to separate truly different clusters far from each other. This nature is called chain effect. A cluster which exists lie intermediately in such a case is called a noise point. However, the clustering using the minimum dissimilarity is robust against existence of outlying values. The concept of the minimum dissimilarity between clusters is shown in Fig. 7.4a.

The maximum dissimilarity is also called a complete link (CLINK), which is equivalent to the longest distance between two clusters, i.e., the distance between the furthest neighbor objects. Although clustering using the maximum dissimilarity is less robust against the existence of outliers than clustering using the degree of minimum dissimilarity, it is easy to make the clustering result compacter and tighter. The concept of the maximum dissimilarity between clusters is shown in Fig. 7.4b.

The average dissimilarity is also called an average link or UPGMA (Unweighted Pair Group Method with Arithmetic Mean). In a word, clustering using the average dissimilarity is characterized between the above two methods for clustering. Therefore the average-based clustering is rather neutral to noise points and outliers. The concept of the average dissimilarity between clusters is shown in Fig. 7.4c.

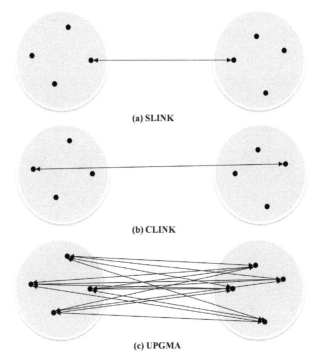

(a) SLINK

(b) CLINK

(c) UPGMA

Figure 7.4 Dissimilarity.

In addition to the above three methods, the Ward method is often used which is based on the distance (Euclidean distance) between the centroids of two clusters as the measure of dissimilarity as follows:

- $$\frac{n_1 n_2}{n_1 + n_2} |m_1 - m_2|^2$$

Here m_1 and m_2 denote the centroids of the clusters c_1 and c_2, respectively.

Notes will be made about chain effects here. They are not necessarily bad. For example, if rather lengthy islands like the Japanese archipelago will be merged to one cluster probably the chain effect will be more suitable than creation of compact and tight clusters. In other words, which type of dissimilarity to choose depends on what kinds of clusters are wanted.

(2) Lance and Williams coefficients

Lance and Williams have shown that the dissimilarity between the newly merged cluster $c_3 (=c_1 \cup c_2)$ and the other clusters (c_1, c_2) can be uniformly calculated by the following formula, using the dissimilarity between the original clusters (c_1, c_2, c_i). This assures that the computation of the dissimilarity can be performed in constant time.

- $d(c_1 \cup c_2, c_i) = a_1\, d(c_1, c_i) + a_2\, d(c_2, c_i) + b\, d(c_1, c_2) + c|d(c_1, c_i) - d(c_2, c_i)|$

Here the four coefficients a_1, a_2, b, and c in the above formula are determined according to the kinds of dissimilarity to use as shown in Fig. 7.5.

	a_1	a_2	b	c
SLINK	1/2	1/2	0	-1/2
CLINK	1/2	1/2	0	1/2
UPGMA	$\dfrac{n_1}{n_1 + n_2}$	$\dfrac{n_2}{n_1 + n_2}$	0	0
Ward	$\dfrac{n_1 + n_i}{n_1 + n_2 + n_i}$	$\dfrac{n_2 + n_i}{n_1 + n_2 + n_i}$	$-\dfrac{n_i}{n_1 + n_2 + n_i}$	0

Figure 7.5 Lance and Williams coefficients.

(3) HDC

HDC is done by the following algorithm. Let the number of clusters desired be k ($<= n$).

(Algorithm) HDC

1. Start from the cluster set C which consists of only one cluster containing all the objects /* At this time $|C| == 1$;
2. Repeat the following procedure while the given condition (usually, $|C| < k$) is true {
3. Choose a cluster c from C and delete it from C;
4. Divide C into l partitions (usually, $l = 2$) according to a certain principle (for example, divide at the place where the distance between the most similar objects is maximized);
5. Let such clusters be the cluster c_i ($i = 1, ..., l$) and add them to C}

The concept of HDC for 2 divisions is shown in Fig. 7.6.

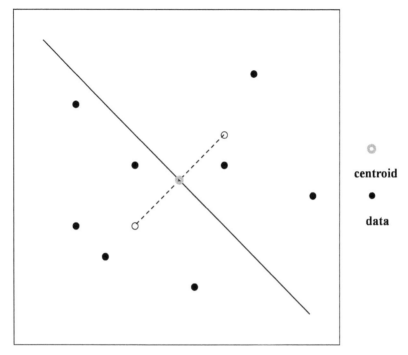

Figure 7.6 HDC.

centroid

data

7.7 Evaluation of Clustering Results

After clustering is performed, it is necessary to evaluate the result. Inherently, however, there exists no correct answer as to what should be returned as a result of clustering, as already described. It is essentially difficult to evaluate the quality of resultant clusters for that reason. Nevertheless, some evaluation measures will be introduced.

(1) Evaluation by the analyst

The user (especially analyst) evaluates the clustering result and gives marks to it by inspection. Since the evaluation by humans tends to easily be subjective, it is sometimes necessary to use and equalize two or more users' evaluations.

(2) Entropy

Although it is not generally possible, if the category (i.e., class) to which each data should belong is sometimes known beforehand, each category is assumed to correspond to a separate cluster, respectively. That is, if there are k categories, it can be assumed that k clusters are made as a result of clustering. Entropy can be used as an evaluation measure under such a premise.

The entropy about the cluster D_i is calculated by the following formula.

- $entropy(D_i) = -\sum_{j=1}^{k} p_i(C_j) \log p_i(C_j)$

Here $p_i(C_j)$ denotes the ratio of data with C_j as its category over all the data contained in the cluster D_i. The entropy of the whole cluster is calculated as follows.

- $entropy_{total}(D) = \sum_{i=1}^{k} \frac{|D_i|}{|D|} entropy(D_i)$

(3) Purity

Purity is also a measure used under the same premise as entropy. The purity about the cluster D_i is calculated by the following formula.

- $purity(D_i) = max_j(p_i(C_j))$

Furthermore, the purity of the whole cluster is calculated like entropy as follows.

- $purity_{total}(D) = \sum_{i=1}^{k} \frac{|D_i|}{|D|} purity(D_i)$

(4) Internal measure

While the measures such as entropy and purity use external data as well, another measure uses only the given data. The degree of cohesion of a cluster can be measured using the squared error mentioned already. Moreover, the degree of separation between two clusters can be measured using the distance between the centroids of the two clusters.

References

[Han et al. 2006] J. Han and M. Kamber: Data Mining: Concepts and Techniques, Second Edition, Morgan Kaufmann (2006).
[Tan et al. 2005] P.-N. Tan, M. Steinbach and V. Kumar: Introduction to Data Mining, Addison-Wesley (2005).

Classification

First, in classification, on the basis of data whose classes or categories are known in advance, classifiers, or mechanisms for correctly assigning appropriate classes to the data, such as classification rules are learned. Second, if new data are given, they are classified by using the learned classifiers. This chapter will describe the method for construction of such classifiers.

8.1 Motivation

When there is a customer application for a credit card, whether to issue a credit card to the customer or not is an important problem for the credit card company. This business is called credit business. From the data related to customers in the past, credit card companies have learned the rules of decision as to whether to issue a card to a new customer, in other words, which conditions such a customer should satisfy. To learn the rules of decision by using data samples of the past and to determine whether 'yes' or 'no' as to new data or assign appropriate classes to new data, is classification [Mitchell 1997, Witten et al. 1999, Han et al. 2001, Hand et al. 2001]. In particular, rules which are used to make a decision at this time are called classification rules. In other words, it is a prerequisite for classification that a class (or category) to which data should belong is known in advance. As already described, the clustering task for partitioning data into groups according to the degree of similarity is significantly different from the classification task because the characteristics and the names of groups may be unknown in advance in the former task.

Classification where the result of determination is a continuous numeric value instead of a discrete value (i.e., class) is especially called prediction or regression. Prediction will be described in a separate chapter.

8.2 Classification Task

This section describes the steps for classification together with the basic concept of classification. Classification consists of two steps: The learning step and classification step (in a narrow sense). Let's consider a database consisting of a set of tuples here. More generally, each tuple is assumed to be described by multiple attributes. In the learning step, a classifier is learned by using the sample data (i.e., tuples). In general, knowledge to be found or patterns or relationships to be learned by data mining is called model. The models include association rules (i.e., frequent patterns) and clustering results in addition to classifiers. In the classification step, newly emerged data are classified by using a learned model. Those two steps will be explained in detail below.

(1) Learning step

A classifier as a model is learned or constructed so as to describe data whose classes are already determined. Each tuple in the database has a class attribute (target attribute) and stores the class name as its value. A tuple with a certain class as a value of its class attribute is interpreted to belong to that class. The model is constructed by analyzing a set of tuples. In that sense, a set of sample data used to construct the model is called training set.

Each sample is given its class name in advance. In the sense that correct classification is provided in advance, classification is called *supervised* learning. On the other hand, clustering is called *unsupervised* learning in that classes of data are not known in advance. Models thus learned are represented by decision trees, rules, or formulas.

(2) Classification step

New data are classified by using a model (i.e., a classifier) obtained by learning. Now let us define the accuracy of the model. So as to determine the accuracy of the model, the data set, called test set, which consists of samples attached with correct class names is used. The accuracy of the model is the ratio of correct predictions over the total number of predictions by the model using the test set. The test set is usually different from the training set. If the model has a precision that can be accepted, then the model is used to classify new data. Note that if a model is selected by evaluating the accuracy based on only the training set, then the model tends to be over-fit. Such evaluation is called optimistic evaluation.

While to predict the value of the class attribute of a discrete type, where the number of distinctive values (i.e., class labels) is relatively small, is called classification, to predict the value of the attribute of a continuous type is called regression. Applications of classification range from those targeting structured data such as direct marketing and credit business to those targeting relatively unstructured data such as classification of social data and Web documents.

8.3 Induction of Decision Tree

A decision tree is often used as a classifier. The decision tree is a tree structure like a flow chart. The non-leaf node of the decision tree represents the conditional test for an attribute and each branch to the child node represents the corresponding result of the test while the leaf node represents the determined class.

Now let's consider classification of unknown samples by using a decision tree. The path from the root node to the leaf node corresponds to a process of determining a class and the leaf node at the end of the path holds the determined class at that time. The decision tree can be converted to classification rules in a straightforward manner. In a word, all the tests along the path connected with logical "AND" and the class represented by the leaf node at the end of the path correspond to the precondition of a classification rule and the conclusion of the rule, respectively.

- The algorithm for induction of decision trees

The induction algorithm of the decision tree will be described below. It is a basic algorithm called ID3 of Quinlan [Mitchell 1997, Han et al. 2001]. This algorithm assumes that category attributes are of a discrete type. Therefore, if the algorithm is applied to a numerical attribute, it is necessary to discretize the attribute values as is the case in association rules.

(Algorithm) Decision tree induction
Input: Training data and attribute list
Output: Decision tree

1. Create a single node N for the samples in the training data;
2. If all the samples belong to the same class, let the node N be a leaf node and label the leaf node with the class name and terminate;
3. If the attribute list is empty, let the node N be a leaf node and label the leaf node with either a default class or the most common class and terminate;
4. Select the test attribute by using a certain measure (e.g., information gain) which can best divide samples into classes;
5. Label the node N with the selected test attribute;
6. Do the following procedure for each value a_i of the test attributes {
7. Create a branch (test attribute $= a_i$) from the node N;
8. Let s_i be a subset of the sample data that satisfy the branch condition;
9. If S_i is empty, attach the branch to a leaf node that is labeled with either a default class or the most common class;

10. Otherwise, let s_i and {the attribute list minus the test attribute} be new training data and a new attribute list, respectively, apply the algorithm recursively, and attach to the branch the decision tree returned as a result of the recursive application;};

8.4 Measure for Attribute Selection

Here, a measure used in selecting the appropriate attributes in the algorithm for decision tree induction will be described. One of the frequently used measures is a measure called information gain or entropy reduction. An attribute that maximizes the value of the measure is selected as a test attribute.

The samples S are assumed to consist of s sample pieces and each sample is assumed to consist of r attributes. The class attribute of the sample is assumed to have any class label representing the class C_i ($i = 1, m$). C_i is assumed to contain s_i pieces as follows.

- $S = \cup C_i, s = |S|, s_i = |C_i|$ ($i = 1, m$)

The probability that a sample belongs to C_i is expressed as follows.

- $p_i = s_i/s$

The expected entropy is expressed by the following equation.

- $I(s_1, s_2, ..., s_m) = -\sum_{i=1}^{m} p_i \log_2 p_i$

On the other hand, assuming that the attribute A has a distinctive value a_j ($j = 1, v$), S is alternatively expressed as follows.

- $S = \cup S_j$ ($j = 1, v$)

Here S_j is a subset of S that satisfies the condition that $A = a_j$. S_{ij} is samples belonging to the class C_i among S_j. If A is selected as the test attribute, then each subset is the branch from A.

The entropy based on the division by the attribute A is calculated as follows.

- $E(A) = \sum_{j=1}^{v} (\dfrac{\sum_{i=1}^{m} s_{ij}}{s}) \cdot I(s_{1j}, s_{2j}, ..., s_{mj})$

Here the entropy for each subset S_j is expressed as follows.

- $I(s_{1j}, s_{2j}, ..., s_{mj}) = -\sum_{i=1}^{m} p_{ij} \log_2 p_{ij}$

Here $p_{ij} = s_{ij}/|S_j|$, which represents the probability of samples belonging to C_i among the samples of S_j.

The information gain can be defined by using the above formulas as follows.

- $Gain(A) = I(s_1, s_2, \ldots, s_m) - E(A)$

This measure corresponds to the entropy which can be reduced by knowing the value of the attribute A.

The information gain is calculated for each attribute and then an attribute with the maximum value is selected as the test attribute for S. That is, a new node is created and labeled with the test attribute and then a branch is created so as to correspond to samples with the same attribute value.

For example, let's consider the training set as illustrated in Fig. 8.1. Letting s_1 and s_2 be YES and NO for wine purchase, respectively, the entropy is calculated as follows:

- $$I(s_1, s_2) = -\frac{9}{14}\log_2\frac{9}{14} - \frac{5}{14}\log_2\frac{5}{14} = 0.940$$

Next let's consider the division by the travel attribute. The entropy for "travel = like" is calculated as follows.

- $$I(s_{11}, s_{21}) = -\frac{6}{8}\log_2\frac{6}{8} - \frac{2}{8}\log_2\frac{2}{8} = 0.811$$

RID	Age	Annual income	Sex	Travel	(CLASS) *Purchase wine*
1	<30	high	male	like	no
2	<30	high	male	dislike	no
3	[30, 39]	high	male	like	yes
4	≥40	medium	male	like	yes
5	≥40	low	female	like	yes
6	≥40	low	female	dislike	no
7	[30, 39]	low	female	dislike	yes
8	<30	medium	male	like	no
9	<30	low	female	like	yes
10	≥40	medium	female	like	yes
11	<30	medium	female	dislike	yes
12	[30, 39]	medium	male	dislike	yes
13	[30, 39]	high	female	like	yes
14	≥40	medium	male	dislike	no

Figure 8.1 Training data.

The entropy for "travel = dislike" is calculated as follows.

- $I(s_{12}, s_{22}) = -\dfrac{3}{6}\log_2\dfrac{3}{6} - \dfrac{3}{6}\log_2\dfrac{3}{6} = 1.00$

Therefore the entropy for the division by the travel attribute is calculated as follows.

- $E(travel) = -\dfrac{8}{14}I(s_{11}, s_{21}) - \dfrac{6}{14}I(s_{12}, s_{22}) = 0.892$

After all, the information gain in this case is calculated as follows.

- $Gain(travel) = I(s_1, s_2) - E(travel) = 0.048$

The information gain is also calculated for each attribute other than the travel attribute as follows.

- *Gain* (age) = 0.246
- *Gain* (income) = 0.029
- *Gain* (sex) = 0.151

By comparing the value of the information gain calculated for each attribute, the age attribute with the largest value is selected as the test attribute. The same process is repeatedly done for each subtree. The Fig. 8.2 illustrates a decision tree that is learned by the ID3 algorithm based on the information gain measure.

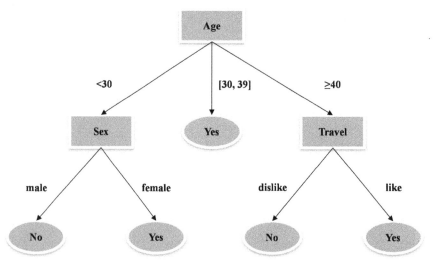

Figure 8.2 Decision tree for "Purchase wine".

8.5 Creation of Classification Rules

Once a decision tree is induced, classification rules can be created in a straightforward manner by using the decision tree. Let's consider one path leading to the leaf node from the root. Then the pair <attribute, attribute value> corresponding to the non-leaf node becomes a simple condition. Simple conditions which are connected by logical "AND" along the whole path is the condition part of the rule. On the other hand, the leaf node is the conclusion part of the rule. For example, using the decision tree induced now, the following rules will be formulated.

IF Age < 30 AND Sex = female THEN Purchase wine = Yes
IF 31 <= Age <= 39 THEN Purchase wine = Yes

8.6 Extension to the Basic Algorithm

For the basic ID3 algorithm, some extensions such as C4.5 [Mitchell 1997, Han et al. 2001] have been considered. Although the detailed explanations will be omitted, such extensions include the following.

- Extension to the prediction of continuous values such as regression trees
- Use of alternative measures for attribute selection such as GINI and AIC other than information gain
- Completion of missing values

However, there still remain other issues to be solved for the basic or extended algorithms. One of them is scalability. The scalability issues will be described as to classification and other data mining tasks in a separate chapter.

8.7 Model Accuracy

The accuracy of the model will be described here. If the training set is used for both the generation and the accuracy measurement of the decision tree, the estimate may be optimistic, leading to the wrong conclusion using the decision tree. Methods for solving this problem based on sampling have been proposed [Kohavi 1995, Han et al. 2001] as follows.

- Holdout method

First in the Holdout method, sample data are randomly divided into the training set and the test set. The ratio of sizes of the two data sets is usually 2:1. Then a classifier is generated by using the training set and then the accuracy is calculated by separately using the test set.

This method is pessimistic in that it uses only a part of the data to evaluate the accuracy. There is a variation in this method which repeats Holdout for k times by random sampling and measures the accuracy for each round and averages all the values.

- k-fold cross validation

  1. First this method partitions sample data into k non-overlapping subsets of approximately the same size, i.e., $\{S_i\}$ $(i = 1, k)$.
  2. Next the training and testing of the classifier are repeated for k times by using the following setup.
  3. {Let a subset S_i of the data and the other subsets $\{S_j\}$ $(j \neq i)$ be the test set and the training set, respectively}.

The final accuracy is determined by dividing the sum of the correct answers for k times by the total number of answers. There is a variation in this method called stratified cross validation which makes the class distribution of each subset substantially the same as that of the original data.

8.8 Accuracy Improvement

In this section, the improvement of the accuracy of the classifier will be explained. In general, there exist general methods such as the Boosting method [Breiman 1996, Witten et al. 1999] and the Bagging method [Freund 1996, Witten et al. 1999]. The basic idea common to these can be expressed as follows.

1. Create a sequence of T classifiers C_i;
2. Construct the final classifier C^* by using C_i.

Here the bagging method will be briefly described in order to explain the above idea.

- Bagging (bootstrap aggregation)

At the time of learning, the Bagging method creates T individual classifiers in sequence. At each iteration (the t-th iteration) in classifier creation, a sample S_t is created by sampling from sample S ($|S| = s$) and then a classifier C_t is created based on S_t. There may exist intersections between distinctive S_t. In reality, each sample is made by deletion and replacement of training data.

In classification of new data, the final classifier C^* takes a majority vote of the classification results of individual classifiers C_i and lets the result be the final result of C^*. The average value is used instead of the majority in the case of numeric attributes.

8.9 Other Models

In addition to the decision tree model, which has been described above, there are various classification methods. Some of them will be briefly introduced below.

- *k*-NN (*k*-nearest neighbor)

k-NN [Mitchell 1997, Hand et al. 2001] takes a majority vote of the classification results of the *k* nearest neighbors to unknown data and lets the result be the final classification result of the unknown data. If the data are represented by *n*-dimensional vectors, *k*-NN often uses the Euclidean distance in the *n*-dimensional space so as to measure the degree of similarity between the data and one of the nearest neighbors.

K-NN doesn't learn until it needs to classify the unknown data, which have no class labels. In that sense, *k*-NN is sometimes called a lazy learner. Thus for the first time at the time of actual classification, *k*-NN makes computation necessary for the class determination. Compared to other classifiers, called eager learners, such as the decision tree, *k*-NN generally needs higher cost at the classification step although *k*-NN needs no cost at the learning step.

Here let's consider the complexity of this method. Assuming that it takes p time to calculate the distance between the query and one piece of data, it takes $O(np)$ as a whole. So indexing is required in order to efficiently select samples. One work of Yu and others [Yu et al. 2001] can be positioned as a research in that direction. K-NN will be touched again in the chapter on Web mining.

- Naïve Bayes

There is a classification method called Naive Bayes based on probability theory [Langley et al. 1992, Mitchell 1997, Han et al. 2001]. The probability $P(X|H)$ that the hypothesis H holds after the event X is observed is called posterior probability. On the other hand, $P(H)$ is called prior probability. For example, the probability that the sample X consisting of n attributes belongs to the class C_i is expressed as $P(C_i|X)$. Assuming the total number of classes is equal to m, the classification problem can be rephrased as follows:

1. Find C_i that maximizes $P(C_i|X)$ $(i = 1, m)$.

From the Bayes theorem, the above probability can be rewritten as follows.

$$P(C_i|X) = P(X|C_i) P(C_i)/P(X)$$

As $P(X)$ is constant for all the classes, $P(X|C_i) P(C_i)$ has only to be maximized so as to maximize this expression. Here, it is estimated that $P(C_i) = s_i/s$. Let s and s_i be the total number of samples and the number of samples included by the C_i, respectively. On the assumption that values of attributes are independent of each other (Naive Bayes assumption), $P(X|C_i)$ can be transformed as follows.

$$P(X|C_i) = \Pi \, P(x_j|C_i)$$

Letting s_{ij} be the number of samples contained by C_i whose $A_j = x_j$, the probability that the value of the category attribute A_j is x_j is equal to $P(x_j|C_i) = s_{ij}/s_i$. Here, it is assumed that $P(x_j|C_i)$ will follow a Gaussian distribution for the continuous values.

For example, let's consider whether a man who is of the age < 30, has low earnings, and dislikes travelling, will buy a wine. Related probabilities are calculated as follows.

> P (Purchase wine = Yes) = 9/14
> P (Purchase wine = No) = 5/14
> P (Age < 30 | Purchase wine = Yes) = 2/9
> P (Age < 30 | Purchase wine = No) = 3/5
> ...
> P (Travel = dislike | Purchase wine = Yes) = 3/9
> P (Travel = dislike | Purchase wine = No) = 3/5

The probabilities to be maximized are calculated by using these values and compared as follows.

> P (Yes) P (< 30 | Yes) P (Low | Yes) P (Male | Yes) P (dislike | Yes) = 0.0053
> P (No) P (< 30 | No) P (Low | No) P (Male | No) P (dislike | No) = 0.0206

Thus, in the Naive Bayes method, "No" is assigned to "purchase a wine" (class attribute) for this example.

As described above, the computational complexity for the Naïve Bayes method can be reduced by assuming independence among attributes. Some methods take dependencies into consideration [Mitchell 1997, Han et al. 2001].

- Support Vector Machine (SVM)

First SVM [Burges 1998] divides the training set into negative examples and positive examples. Next SVM determines the hyperplane between positive and negative data so as to maximize the margin between the positive and negative cases (i.e., the distance between the support vectors corresponding to the forefronts of each case). SVM will be described in more detail in the chapter on Web mining.

References

[Breiman 1996] L. Breiman: Bagging predictors.Machine Learning 24(2): 123–140 (1996).

[Burges 1998] Christopher J.C. Burges: A Tutorial on Support Vector Machines for Pattern Recognition. Data Mining and Knowledge Discovery 2(2): 121–167 (1998).

[Freund 1996] Yoav Freund and Robert E. Schapire: Experiments with a new boosting algorithm. In Proceedings of International Conference on Machine Learning, pp. 148–156 (1996).

[Han et al. 2001] Jiawei Han and Micheline Kamber: Data Mining: Concepts and Techniques, Morgan Kaufmann, August 2001.

[Hand et al. 2001] D. Hand, H. Mannila and P. Smyth: Principles of Data Mining, MIT Press (2001).

[Kohavi 1995] Ron Kohavi: A Study of Cross-Validation and Bootstrap for Accuracy Estimation and Model Selection. In Proceedings of IJCAI, pp. 338–345 (1995).

[Langley et al. 1992] Pat Langley, Wayne Iba and Kevin Thompson: An Analysis of Bayesian Classifiers. In Proceedings of National Conference on Artificial Intelligence, pp. 223–228 (1992).

[Mitchell 1997] Tom M. Mitchell: Machine Learning, McGraw-Hill (1997).

[Witten et al. 1999] Ian H. Witten and Eibe Frank: Data Mining: Practical Machine Learning Tools and Techniques with Java Implementations, Morgan Kaufmann, October 1999.

[Yu et al. 2001] Cui Yu, Beng Chin Ooi, Kian-Lee Tan and H.V. Jagadish: Indexing the Distance: An Efficient Method to KNN Processing. In Proceedings of the 27th International Conference on Very Large Data Bases, pp. 421–430 (2001).

Prediction

Classification described in the previous chapter generally determines discrete categories from given variables. In this section, as a task similar to classification, prediction of continuous variables based on other variables will be explained. In prediction, the former and latter are called dependent variables and independent variables, respectively. As techniques for prediction of dependent variables as functions of independent variables, regression as a basic method and multivariate analysis as its advanced method will be briefly surveyed. Model construction based on these techniques is indispensable so as to create and confirm concrete quantitative hypotheses at the conceptual layer constituting big data application systems.

9.1 Prediction and Classification

Classification and prediction have much in common. Here relations between classification and prediction will be explained first.

In classification, first of all, if data (i.e., records with attributes) and categories (i.e., classes) to which the data belong are provided as samples, then classifiers are learned as a result based on a part or all of them. In actual deployment of classification, if new data whose categories are yet unknown are provided, classes to which they should belong are determined based on the attribute values by using the learned classifiers.

Categories in classification can be considered as equivalent to nominal variables or ordinal variables. In other words, such a categorical variable can be considered as a variable which takes individual categories as discrete values.

Assuming that there is a continuous dependent variable which is determined by independent variables, if concrete values of the dependent variable can be changed into discrete values by certain methods, the dependent variable can be regarded as a kind of categorical variable in some cases. Anyway, classification corresponds to, so to speak, predicting the value of a dependent discrete variable (i.e., a category variable), based on the independent variables, whether they are discrete or continuous.

Indeed, regression trees have been invented as extensions of decision trees which can handle continuous values. However, they can be considered rather as techniques similar to prediction. They determine the value of a dependent variable which generally takes continuous values, based on, whether discrete or continuous, independent variables sticking to trees. If a continuous dependent variable could be discretized by any means, it might be possible to predict the values by applying classification techniques. However, it is generally inappropriate or essentially difficult to discretize continuous variables. In a word, discretization means to express all different continuous values contained in a certain section (i.e., a bin) with the single identifier of the section. Generally speaking, sufficient accuracy cannot be expected for predicted values in case of discretized variables. If the accuracy were to be raised, the required number of the sections would approach that of the distinctive values of the original variables and would become huge, as a result. Moreover, the techniques for classification don't necessarily assume to treat a large number of categories.

Conceptually, it is possible that discrete categorical variables in classification are extended even to continuous variables (i.e., dependent variables) and their values are predicted based on variables (i.e., independent variables) other than the dependent variables. As described above, straightforward application of the techniques for classification to prediction is inappropriate. So techniques suitable for prediction which are different from those used for classifications will be explained below.

Please also note that the concepts about accuracy differ from classification to prediction. Generally in classification, the data used to construct a classification model and the data used to confirm the accuracy of the classification model are prepared separately. In classification, a scheme which uses the same data both for model construction and confirmation is called optimistic and it is desirable that such an optimistic scheme should be avoided. In prediction, on the other hand, the accuracy (more precisely, fit) of a model is calculated based on the differences between observable values used to construct the model and values predicted by the model.

9.2 Prediction Model

In general, independent variables and dependent variables in prediction express causes and effects, respectively. Thus it is possible to predict the values of dependent variables using those of independent variables. Here, it is assumed that both independent variables and dependent variables can be observed. In addition, for the time being, it is assumed that all variables involved in prediction take continuous values.

Here a regression model, which predicts dependent variables with the function of independent variables, will be explained. A case where a single independent variable is involved is called simple regression model while a case where two or more independent variables are involved is called multiple regression model. First, a linear regression model, simply denoted by a linear function, will be explained. Then methods for multivariate analysis such as path-analysis model, multiple indicator model, and factor analysis model will be explained as more advanced models.

9.2.1 Multiple Regression Model

First a multiple regression model which generally predicts one dependent variable by two or more independent variables will be explained. Here, the multiple regression model is explained using a simple example where three independent variables are involved. Assume that a dependent variable (X_4) is expressed as a formula of three independent variables (X_1, X_2, X_3) as follows:

- $X_4 = \alpha_4 + \gamma_{41}X_1 + \gamma_{42}X_2 + \gamma_{43}X_3 + e_4$

Here α_4, $\{\gamma_{41}, \gamma_{42}, \gamma_{43}\}$, and e_4 denote an intercept, partial regression coefficients, and an error, respectively. The intercept is a predicted value in case all the independent variables are 0. The partial regression coefficient of a variable expresses the increase in the predicted value in case the value of the variable is increased by one unit and the other variables are kept as they are. The error is a residual except the observable variables.

The average (expected value) and the variance of each variable can be standardized with zero and one, respectively, without loss of generality. Then α_4 can be set to zero. The expected value of product of each independent variable and the error is also assumed to be zero. Generally, it is assumed that there exists a correlation (i.e., covariance) between two independent variables.

The partial regression coefficient of a variable is determined so that the sum of squares of the difference (i.e., error) between the observed value and predicted value of the variable becomes as small as possible. The coefficient of determination will be introduced here. It is calculated as the square of the multiple correlation coefficient R between the observed value and predicted value of X_4. The coefficient of determination, denoted by R^2, is defined as follows:

$$R^2 = 1 - \frac{\Sigma(observed\ value - predicted\ value)^2}{\Sigma(observed\ value - average\ of\ observed\ values)^2}$$

The meaning of the multiple correlation coefficient can be interpreted as the independent variable's contributions to prediction of the dependent variable. Moreover, there are several indexes as fit of a model. For example, a *chi* square statistic is often used among them. This measures goodness of fit by the discrepancy between the observed value and predicted value. There are many indexes of model fit such as GFI in addition to a chi square statistic.

A multiple regression model will be explained by using a fictitious example. Let us consider predicting the safety of a railway vehicle as a dependent variable by the mileage, the use years, and the months until next legal inspection, as independent variables, assuming that the safety can be calculated by the rate of incidents. The model is illustrated in the Fig. 9.1.

In the figure, R^2 indicates the coefficient of determination of the model and the values on the paths from each of the independent variables to the dependent variable such as $\gamma_{\text{safety use-years}}$ indicate the partial regression coefficients.

Figure 9.1 Multiple regression model.

9.2.2 Transformation of Nonlinear Functions

Here the relation between a multiple regression model and a simple regression model will be explained from an angle other than the number of independent variables. Of course, the simple regression model corresponds to the reduced multiple regression model involving only a single independent variable. Therefore, generally the simple regression model is expressed as follows:

- $Y = cX + e$

However, a nonlinear function of an independent variable may be able to more accurately predict a dependent variable than a linear function, depending on applications of a simple regression model. A case where a dependent variable is predicted with a nonlinear function of one independent variable will be considered below. For example, let us consider predicting a dependent variable using the 3rd order polynomial function of an independent variable as a kind of nonlinear function as follows:

- $Y = aX^3 + bX^2 + cX + e$

By using our knowledge about the multiple regression model, let us consider that the above polynomial function will be transformed. If three new variables are introduced as $X_3 = X^3$, $X_2 = X^2$, and $X_1 = X$, respectively, the above formula will become the multiple regression model containing the three independent variables X_1, X_2, and X_3. In general, by introducing two or more new independent variables instead of one independent variable, the simple regression model originally denoted by the nonlinear polynomial of one independent variable can be transformed into the multiple linear regression model which consists of two or more independent variables. Thus, the dependent variable can be predicted by the independent variable using the multiple regression model constructed in this way. Generally in a case of the *p*-th order polynomial, more than *p* distinctive data are needed. In particular, only in cases where the value of an independent variable changes within a small range and the order *p* is small, prediction by applications of this technique is recommended.

9.2.3 Path Analysis Model

Let us take the example used in the subsection about multiple linear regression analysis again. If this example is more deeply considered, the *use years* affect the *mileage* with high possibility. Moreover, it is expected that there are no relations between the *use years* and the remaining *months until* the next legal *inspection*. If such conditions are taken into consideration, this example will exceed the capability of multiple regression models. That is because multiple regression models generally assume that there exists only one dependent variable which is affected by independent variables and there exist correlations between independent variables. However, the existence of "too strong" correlations between independent variables incurs another problem called multicollinearity.

One of the analysis models which can treat such constraints directly, is path analysis model [Kline 2011]. The path analysis model representing this example is illustrated in Fig. 9.2.

In the path analysis model, each value attached to the path such as $\gamma_{safety\ use\text{-}years}$ is called a path coefficient, which is equivalent to the partial regression coefficient in the multiple regression model (see Fig. 9.2). The average and variance of a variable are set to zero and one, respectively, by standardization similarly as previously.

In this example, the *use years* affect the *safety* through two paths. One direct path corresponds to the direct effect on the safety by the use years

Figure 9.2 Path analysis model.

while another indirect path through the mileage corresponds to the indirect effect. Product of all the coefficients on the indirect path accumulates the indirect effect. Sum of the effects of all the paths which originate in the independent variable *use years* and reach the dependent variable *safety* accumulates the contribution as a whole to the dependent variable by the independent variable, i.e., the total effect from the *use years* to the *safety*. That is, the total effect in this case is calculated using all path coefficients as follows:

$$\gamma_{\text{mileage use-years}} \times \gamma_{\text{safety mileage}} + \gamma_{\text{safety use-years}}$$

9.2.4 Multiple Indicator Model

So far it has been assumed that all variables can be observed. However, all variables cannot always be observed while both independent variables and dependent variables can be observed. Some applications need to treat abstract concepts, such as intelligence and popularity. It is not possible to know what such abstract concepts can take as their values nor to observe the values. The techniques invented so as to consider such latent structures as targets of analysis are collectively called covariance structure analysis or Structural Equation Modeling (SEM), which includes a factor analysis model and a multiple indicator model. Here the variable which is not observed is called a latent variable.

First the multiple indicator model will be described. The multiple indicator model assumes that there exist causal relationships between latent variables. The multiple indicator model containing two latent variables (F_1 and F_2) representing two factors is shown in Fig. 9.3. Here the factor F_1 affects the factor F_2.

$X_1 = \lambda_{11}F_1 + e_1$: measurement equation
$X_2 = \lambda_{21}F_1 + e_2$
$X_3 = \lambda_{32}F_2 + e_3$

$$X_4 = \lambda_{42}F_2 + e_4$$
$$F_2 = \gamma_{21}F_1 + d_2: \text{structural equation}$$

$\lambda_{11}, \lambda_{21}, \lambda_{32}, \lambda_{42}, \gamma_{12}$ are equivalent to partial regression coefficients in the multiple linear regression model and also called path coefficients as in the path-analysis model. The equations involving observable variables and the equations concerning only latent variables are called measurement equations and structural equations, respectively. However, both of them have completely the same structures. The expected value and variance of each factor are standardized to zero and one, respectively. Moreover, causal relationships are assumed to exist between factors.

e_i and d_2 cannot be observed. In that sense, they are also a kind of latent variables. They are, so to speak, errors. Furthermore, it is assumed that there exist no correlations between two distinctive elements of the errors $\{e_i, d_2\}$ nor between any factor and any error.

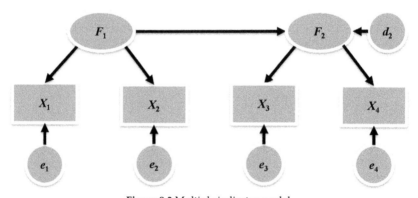

Figure 9.3 Multiple indicator model.

9.2.5 Factor Analysis Model

The factor analysis model considers two or more latent variables like the multiple indicator model. A latent variable in the factor analysis model is called a common factor or shortly a factor. The factor analysis model generally presumes that observed variables can be explained by two or more latent variables.

On the other hand, unlike the multiple indicator model, the factor analysis model allows no causal relationships between latent variables (i.e., factors) and none between observed variables. Generally, however, the factor analysis model assumes correlations between factors and ones between observed variables.

In the factor analysis model, factors as common causes induce correlations between two or more variables. For example, the simple model consisting of four observed variables and two factors is expressed as follows (see Fig. 9.4).

$$X_1 = \lambda_{11}F_1 + \lambda_{12}F_2 + e_1$$
$$X_2 = \lambda_{21}F_1 + \lambda_{22}F_2 + e_2$$
$$X_3 = \lambda_{31}F_1 + \lambda_{32}F_2 + e_3$$
$$X_4 = \lambda_{41}F_1 + \lambda_{42}F_2 + e_4$$

λ_{ij}, which have been called path coefficients so far, are especially called factor loads in the factor analysis model. In this example, each observed variable is affected from both of the two factors.

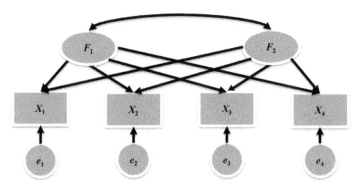

Figure 9.4 Factor analysis model.

9.2.6 Rotation of Factors

As already explained, there are exploratory data analysis as a task for hypothesis construction and confirmatory data analysis as a task for hypothesis verification. Especially, in factor analysis, the former and the latter are called explanatory factor analysis and confirmatory factor analysis, respectively. In confirmatory factor analysis, among factor loads, which represent influences from factors, some factor loads are fixed to constants (usually zero) and the remaining factor loads are presumed as variable parameters. On the other hand, explanatory factor analysis determines all the factor loads without selecting specific factor loads and fixing the values to constants as confirmatory factor analysis.

Here, the explanatory factor analysis model is considered.

First two or more factors are assumed to be a group of basis vectors in an orthogonal coordinate system. This assumption is realized if the factors have no correlation mutually. Then, each variable can be considered a vector in this orthogonal coordinate system which has factor loads as components. If factor loads (i.e., components of vectors) can be changed so that the lengths and relative spatial relations between vectors remain the same even if this orthogonal coordinate system is rotated, it can be said that the set of the vectors represent the set of the original variables.

Now, by using an arbitrary orthogonal matrix, i.e., a square matrix S whose transposed matrix is its inverse matrix, the relations between variables and factors can be transformed as follows:

$$x = \Lambda f + e = (\Lambda S^T)(Sf) + e$$

The vectors x, f, and e express observed variables, factors, and errors, respectively. The matrix Λ expresses factor loads. Here if Sf and ΛS^T are taken as new factors and new factor loads, respectively, the values of observed variables remain the same.

First of all, an orthogonal matrix generally expresses the rotation of a vector. That is, rotation of a coordinate system and adjustment of factor loads can be done by an orthogonal matrix. Generally if two vectors do not change a relative spatial relation, the correlation coefficient between them does not change as well. Therefore, the matrix consisting of variances and covariances does not change, either. In a word, the explanatory factor analysis model based on covariances (or correlations) essentially has freedom as to rotation.

Next, a case where there are correlations among two or more factors will be considered. In this case, what is necessary is just to consider an oblique coordinate system for transformation, instead of an orthogonal coordinate system. In that case, vectors are rotated using not an orthogonal matrix but a regular matrix.

A regular matrix is a square matrix that has an inverse matrix. Therefore, a regular matrix is also called invertible matrix. That is, even if an arbitrary regular matrix T makes new factors Tf and new factor loads ΛT^{-1}, the variance-covariance matrix still remains the same. As in the case of an orthogonal matrix, if the original factors and factor loads are replaced by new factors and factor loads using a regular matrix, respectively, arbitrary rotations (i.e., oblique rotation) can be performed.

As mentioned above, irrespective of the existence of correlations between factors, the factor analysis model has freedom as to rotation and therefore has a problem such that a model cannot be uniquely determined. As a solution to this model identification problem in factor analysis, a method to choose a transformational matrix that emphasizes differences in the values of factor loads is proposed. That is, the transformational matrix makes loads close to zero closer and loads far from zero further.

9.2.7 Structural Equation Modeling Revisited

The covariance structure model or SEM, which models causal relationships based on variances and covariances between observed variables will be explained again.

Fundamental concepts will be defined first. The average and variance about one variable x are defined as follows:

- $m_x = \dfrac{1}{n}\Sigma_i^n x_i$

- $s_{xx} = \dfrac{1}{n-1}\Sigma_i^n (x_i - m_x)^2$

The covariance of the two variables x and y is expressed like the variance as follows:

- $s_{xy} = \dfrac{1}{n-1}\Sigma_i^n (x_i - m_x)(y_i - m_y)$

Furthermore, the correlation coefficient is defined as follows:

- $r_{xy} = \dfrac{s_{xy}}{\sqrt{s_{xx}\,s_{yy}}}$

Therefore, the fact that the covariance is 0 agrees with the fact that the correlation is 0. That the correlation of two variables is 0 means that the variables are not related to one another, i.e., not correlated. Furthermore, if each variable is standardized so that the variance is 1, the covariance and the correlation coefficient are equal.

The following example is considered again (see Fig. 9.5).

- $X_1 = \lambda_{11}F_1 + e_1$
- $X_2 = \lambda_{21}F_1 + e_2$
- $X_3 = \lambda_{32}F_2 + e_3$
- $X_4 = \lambda_{42}F_2 + e_4$
- $F_2 = \gamma_{21}F_1 + d_2$

A matrix which has variances and covariances as its components will be considered here. Such a matrix is called variance-covariance matrix. The

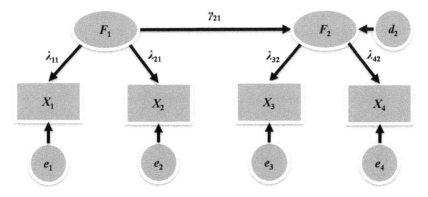

Figure 9.5 SEM.

variance-covariance matrix for the latent variables F_1 and F_2 in the above model is expressed as follows:

- $$\begin{bmatrix} var(F_1) & cov(F_1, F_2) \\ cov(F_2, F_1) & var(F_2) \end{bmatrix}$$

Each component of the matrix is expressed as one path coefficient for the direct path (i.e., direct effect) or product of path coefficients for the indirect path (i.e., indirect effect) between the variables. If there are two or more paths between the variables, the component is expressed as sum of either the path coefficient or products of the path coefficients for each path between the variables (i.e., total effect). Assuming that the latent variables are standardized, some of the components are expressed concretely as follows:

- $var(F_1) = \varphi = 1$
- $conv(F_1, F_2) = \gamma_{21}\varphi = \gamma_{21}$

The variance-covariance matrix for the observed variables X_i is also expressed as follows:

- $$\begin{bmatrix} var(X_1) & cov(X_1, X_2) & cov(X_1, X_3) & cov(X_1, X_4) \\ cov(X_2, X_1) & var(X_2) & cov(X_2, X_3) & cov(X_2, X_4) \\ cov(X_3, X_1) & cov(X_3, X_2) & var(X_3) & cov(X_3, X_4) \\ cov(X_4, X_1) & cov(X_4, X_2) & cov(X_4, X_3) & var(X_4) \end{bmatrix}$$

Similarly some of the components are expressed concretely as follows:

- $var(X_1) = \lambda_{11}^2 + \theta_1 = 1$
- $conv(X_1, X_2) = \lambda_{11}\lambda_{21}$
- $conv(X_1, X_3) = \lambda_{11}\gamma_{21}\lambda_{32}$

Here θ_i denotes the variance of the error e_i.

Thus, the covariance structure model or SEM has a capability to comprehensively express various models such as the multiple regression model, the path analysis model, the multiple indicator model, and the factor model, which have already been described. Please refer to [Kline 2011] for more details about SEM.

9.2.8 Factors Revisited or Dimensional Reduction

The roles of factors introduced by the factor analysis model will be considered again. Factors have been introduced as latent variables which affect observed variables. Structurally speaking, a factor can be viewed as a variable which encapsulates two or more observed variables. Therefore, the factors can reduce the number of the observed variables. Of course, this feature does not necessarily solve the problem of scalability as to computational complexity. However, it can contribute to solving the problem of dimensionality reduction at least at the conceptual level.

Here, data mining techniques useful for reduction of dimensionality will be collectively reviewed. Other scalability-related techniques will be explained in a separate chapter.

By abstracting items conceptually and using items corresponding to super concepts in the concept hierarchy in association analysis, it is possible to substantially reduce dimensions. However, since this generally tends to increase the support counts of itemsets, it may make the throughput (i.e., database access) rather large.

LSI (Latent Semantic Indexing) is often used for clustering and searching text documents. First LSI performs SVD (singular value decomposition). Then LSI chooses the n largest singular values such that n is smaller than the number of dimensions of the original data and embeds the original data into the lower dimensional space by using the dimensions corresponding to the chosen singular values. LSI chooses the dimensions so as to represent the original data as well as possible.

In cluster analysis, LPI (Locality-Preserving Indexing) [Cai et al. 2005] is also useful for reduction of dimensions. This method maps data from higher dimensional space to lower dimensional space, preserving the similarity between data, based on the inner products or cosine measures of data vectors included by k-NN. Both LPI and LSI use SVD in order to remove singular values which are equal to 0. However, LPI primarily aims to preserve distances between data in mapping while LSI aims to represent the original data well in embedding.

However, if the original number of dimensions is N and the number of data is $O(N)$, then the computational complexity of SVD by commonly used methods such as QR is $O(N^3)$. Therefore, in the case of high dimensionality, problems may occur in mere applications of SVD-based methods.

SOM (Self-Organizing Map), which can group similar data like cluster analysis is also useful for data visualization in lower dimensional space. For data of high dimensions at the input layer, SOM finds nodes with weight vectors which have the same number of dimensions with the original data and are the nearest to the data, among the nodes (units) at the output layer, which is a lower dimensional (usually, two or three dimensions) manifold. Such a node is called BMU (Best Matching Unit). Furthermore, SOM updates the weight vectors of both BMU and the nearest nodes to BMU at the output layer so that these nodes are closer to the input data. SOM monotonically decreases both the search range of the nearest nodes of BMU and the incremental value of weights of vectors in each repetition of the above processes. Therefore, in the phase of visualization, similar nodes collected by SOM constitute clusters since each input data are assigned to the nearest node. As for parallel distributed processing of SOM, the author and his colleagues [Goto et al. 2013] used a hashing function which can preserve the proximity among vectors in search of BMU in the environment of MapReduce on Hadoop.

Attribute deletion in classification is to choose only important attributes for classification tasks. It is equivalent to directly performing dimension reduction whether heuristic approaches based on correlation analysis or systematic approaches in target of optimizing accuracy.

Factors in the factor analysis model can be represented by big objects of MiPS model introduced in Chapter 2. In that case, variables which are dependent on factors correspond to attributes of big objects. Factors in the multiple indicator model can be similarly represented by big objects. Thus, the curse of high dimensionality at the attribute level can be lowered at least at the conceptual level of big objects.

References

[Cai et al. 2005] Deng Cai, Xiaofei He and Jiawei Han: Document Clustering Using Locality Preserving Indexing, Ieee Transactions on Knowledge and Data Engineering 17(12): 1624–1637 (2005).

[Goto et al. 2013] Yasumichi Goto, Ryuhei Yamada, Yukio Yamamoto, Shohei Yokoyama and Hiroshi Ishikawa: SOM-based Visualization for Classifying Large-scale Sensing Data of Moonquakes, In Proc. 4th International Workshop on Streaming Media Delivery and Management Systems, Compiegne France (2013).

[Kline 2011] R.B. Kline: Principles and practice of structural equation modeling, Guilford Press (2011).

Web Structure Mining

This chapter explains basic concepts about Web mining and focuses on Web structure mining. After bibliometrics as a preliminary stage of Web structure mining is introduced, methods for computing values of academic researchers and Web pages are described as Web structure mining.

10.1 Web Mining

Data-intensive Web systems typically consist of Web contents (i.e., Web pages), Web access logs (i.e., user access histories) in a Web server and databases at its back end (see Fig. 10.1). In other words, such a Web system as a whole constitutes a Web database system. Among such data, Web contents are modeled as graph structures where pages and links (hyperlinks) correspond to nodes and edges, respectively. Web contents in a narrower sense may represent multimedia data such as texts and photos within the Web pages except the links. The author takes this position.

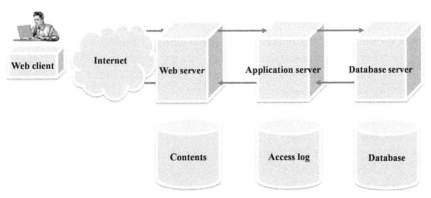

Figure 10.1 An architecture for Web database.

Here let the targets of Web mining be links, texts in pages, and access logs. Mining multimedia data and databases will be described separately. Therefore Web mining is roughly classified into the following three categories according to its main target:

1. Structure mining targets graph structures of Web pages (i.e., link structures).
2. Contents mining targets contents of Web pages (i.e., texts).
3. History mining targets Web access logs.

Please note that some researches or technologies may not be strictly classified into only one category.

In this section, technologies which find meaningful patterns or structures by paying attention to the graph structures of Web pages will be explained in order as follows:

- bibliometrics (an impact factor and h-index)
- Web link analysis (prestige, PageRank, and HITS)

10.2 Structure Mining

10.2.1 Bibliometrics

Bibliometrics has been a scientific field since before the Web emerged and it aims at identifying influential writings (especially academic books and papers) and authors and also the relationships between them through quantitative analysis of writings and authors. Bibliometrics has invented at least the following concepts and laws until now.

- The law of Lotka is a statistic law about the writing productivity of an author.
- The law of Zipf is a statistic law about the contents of writings.
- The number of times that a certain writing is cited by another writing is deeply related to the influence of the cited writing.
- Co-citation means that two or more writings simultaneously cited by another writing (i.e., two or more writings which coincide in citation) can be used for measuring similarity between the cited writings in the analysis.
- Co-reference means that two or more writings citing another writing in common can be used for measuring similarity between the citing writings in that case.
- An impact factor is calculated by analyzing the times of citations of writings published at the academic journal and can be used for measuring the influence of the journal based on the results.

If writings and citations are extended to pages and links, respectively, the above laws and concepts can be used for analysis of the Web as well as writings. First the laws of Lotka and of Zipf will be explained briefly. The remaining concepts relevant to citations will be described in detail later at appropriate places.

(1) The law of Lotka

This is a statistic law about the frequency distribution of authors' productivity. Let P be the number of the writings which an author published and let A be the frequency of such an author, then the following empirical rule holds:

- $A \propto P^{-c}$

Here, c is a positive number (around 2). The law of Lotka insists that the more writings an author publishes, the smaller the frequency of such an author is (see Fig. 10.2a).

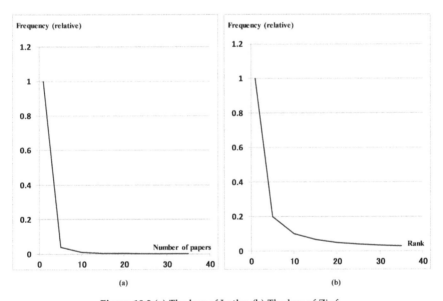

(a)   (b)

Figure 10.2 (a) The law of Lotka, (b) The law of Zipf.

(2) The law of Zipf

This is a statistic law about the frequency distribution of words which appear in a writing. Let R be the rank of a word used in the writing and let W be the frequency, then the following empirical rule holds.

- $W \propto R^{-1}$

The law of Zipf insists that the frequency of a word is in inverse proportion to the rank of the word (see Fig. 10.2b). The laws of Lotka and of Zipf are kinds of power laws [Broder et al. 2000].

10.2.2 Cited Reference Database and Impact Factor

(1) Cited Reference Database

In relation to citations, the databases of cited references which can be used on the Web will be explained. Google Scholar [Google Scholar 2014] and CiteSeer^X [CiteSeer^X 2014] are Web Services as to cited reference databases which allow the users to find papers citing a certain paper and to know the number of citations of the paper.

Google Scholar, one of the services which Google provides, enables the users to retrieve literature information by specifying search terms such as an author, a title, keywords, and publication time, if required. Information such as the title, author, source, electronic edition (PDF and PS forms), papers which cite it, places where they cite it, and the number of citations are returned as search results. Thus, the pages and links correspond to the papers and the citation relationships between the papers, respectively.

Generally these cited reference databases are automatically built based on the data on the Web. That is, papers are collected by using information as to the programs of international conferences or the publication of journals, which are acquired by execution of search engines, monitoring of mailing lists and posting sites, and direct access to sites of publishing companies. Analysis of the collected literatures is also conducted automatically.

DBLP (Digital Bibliography & Library Project) [DBLP 2014] is similar to Google Scholar. The DBLP service, which manages about 1.6 M articles of the computer science field, is founded by Michael Ley, a university researcher. Papers are listed in reverse chronological order for each author. Each paper entry includes the title, source, and the links to pages where its detailed bibliographic information and its electronic editions can be obtained. The entry also includes links to pages about the paper in similar services such as Google Scholar and CiteSeer^X which allow the users to know the citation relationships of the papers. Of course, papers can be searched by specifying a conference or a journal as well as an author or a title. A page for each author in DBLP is also created automatically. Table-of-contents (TOC) pages of journals or international conferences are made first and they are stored into one file called TOC OUT. From TOC OUT file, pages for authors are created and the names of all authors are extracted and stored into one file called AUTHORS. Furthermore, links to pages for authors are embedded into TOC pages using TOC OUT and AUTHORS files.

Moreover, ArnetMiner, a Web Service founded as a research project of the Tsinghua University [ArnetMiner 2014], exhibits the results of mining the social networks which consist of researchers (i.e., authors), conferences, and writings. In ArnetMiner, an author page includes the author's profile and list of writings, the citation of the writings as well as the author's h-index, which will be explained later. In addition, the ranks of the writings, the conferences and the journals are included by the ArnetMiner pages.

Furthermore, Web of Science is among the commercial cited reference databases [Web of Science 2014]. Science Citation Index (SCI) provided by the database has some authority as indices for measuring researchers' achievements in that the collection of analyzed journals is controlled. First a basic set of journals is considered and is also called SCI. Then another set of journals (8608 journals as of this writing) called expanded SCI is considered as a superset of SCI and the database of cited references is created by using expanded SCI. That is, the number of citations of papers is calculated based on expanded SCI.

For example, SCI includes the following journals in fields related to database and data mining.

- ACM TRANSACTIONS ON DATABASE SYSTEMS (TODS)
- IEEE TRANSACTIONS ON KNOWLEDGE AND DATA ENGINEERING (TKDE)

However, details as to how SCI is constructed are unknown since it is commercial.

In Japan, the National Institute of Informatics (NII) operates the NII cited reference information navigator CiNii [CiNii 2014]. While Google is a general-purpose search engine, Google Scholar is a vertical search engine, which specializes in search of papers in academic fields such as computer science.

(2) Impact factors

Impact factors will be explained in relation to citation. An impact factor of a certain journal with respect to a specific year is calculated as summation of the numbers of citations in the specified year, of all the papers published on the journal in the preceding two years divided by the total number of the papers. That is, let *<number-of-citations yp>* be the number of citations of a paper *p* in a certain year *y*, *<impact factor y>* be the impact factor of the journal of the year *y*, and *{paper y–2, y–1}* be a set of the papers published during the preceding two years. Then, the impact factor will be defined by the following formula:

(Definition) Impact factor

- $impact\ factor\ y = \dfrac{\Sigma_{p \in \{paper_{\ y-2,y-1}\}} number\text{-}of\text{-}citations_{yp}}{|\{paper_{\ y-2,y-1}\}|}$

In a word, the impact factor of a journal as to a year represents the average number of citations of all the papers published in the journal for the preceding two years.

For example, the impact factors for 2009 of TODS and TKDE in computer science are 1.245 and 2.285, respectively. In general, journals with high impact factors can be considered as important. NATURE and SCIENCE are among the top journals in the domain of natural science and the impact factors for 2009 are 34.480 and 29.747, respectively. The impact factors of these scientific journals are higher than those of the above journals in computer science.

Please note that the average number of pages of papers differs greatly depending on the journals. This difference may be excessively large among different fields (for example, natural science and computer science). The typical paper lengths of NATURE and SCIENCE are around 2 and 4, respectively. On the other hand, some papers appearing in TODS may amount to 50 pages. If highly cited papers are contained in a journal, the impact factor of the journal is naturally raised. Especially, if a journal contains review papers (i.e., survey papers), the impact factor of the journal will tend to increase. Moreover, the impact factors cannot reflect situations of fields where papers published three or more years ago are frequently cited, or the half-life of citations, so to speak, is long. Anyway, although the impact factor is definitely one index of measuring the influences of journals, it cannot be used for measuring the importance of individual papers, for the above reasons.

10.2.3 H-index, or Value of an Academic Researcher

How is the value of a researcher, who is expected to publish academic papers, assessed? Of course, the whole value of a researcher cannot be represented only with the authored papers. In addition to research competency itself, the capabilities of managing research projects and of educating students and contributions to academic communities, industries, and societies in general are actually required of researchers. Nevertheless, let us consider the value of a researcher only from a viewpoint of paper authoring. For example, is the average of the numbers of citations of all the papers that the researcher writes appropriate for deciding the value? Or should the maximum value of the numbers of citations be used instead?

Use of the average number of citations may be disadvantageous for researchers who produce a lot of papers while its use may be advantageous for researchers who produce only a few papers. Moreover, only the maximum number of citations of researchers cannot reflect the productivity of the researchers.

Jorge E. Hirsch, a physicist, has proposed h-index as one answer to these problems [Hirsch 2005]. That N as a value of h-index of a researcher implies that at least N papers of the researcher have been cited at least N times. Unlike methods based on aggregate functions such as average and maximum, the h-index of a researcher, just a scalar value, can represent both the productivity and the prestige of a researcher with respect to the papers that the researcher wrote. Conversely, h-index cannot express the extremes value such as the maximum number of citations or the total number of written papers. Needless to say, the h-index as well as the impact factor is not suitable for comparison of the values of researchers from different fields.

Some academic services such as Publish or Perish, scHolar index, calculate the h-index using the number of citations provided by Google Scholar, which is a specialized search engine. However, the method itself for calculating each researcher's h-index is very straightforward. On top of the graph (see Fig. 10.3) where the x-axis and y-axis indicate the rank of a paper of a researcher and the number of citations of the paper, respectively, a straight line of y = x is drawn. If we search for the largest rank of a paper whose number of citations lies on or above the straight line, it will be the h-index of the researcher. If the value of the h-index is squared, then the order of the total number of citations as to the researcher can be estimated.

Incidentally, the article on predicting Nobel laureates based on the scheme of PageRank was published in arXiv, an open access academic journal. According to the article [Maslov et al. 2009], the authors have adapted the PageRank algorithm to the citation networks of papers published in physics journals such as "Physical Review Letters" after

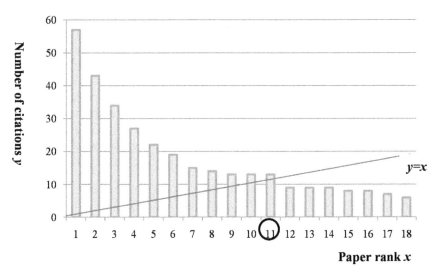

Figure 10.3 Paper rank vs. number of citations.

1893 and have calculated the ranks of the papers. They have discovered the fact that authors of the top 10 papers won the Nobel Prizes and have conjectured that future Nobel laureates may be predictable if this scheme could be applied to more newly published papers. However, since citations of the papers published after the Nobel Prize awards are also included in the calculated ranks, the scheme will require further improvement.

10.2.4 Prestige

The concept of prestige [Liu 2007] of an actor in social network analysis can be modeled by using graphs as the data structures. Bibliometrics considers papers as such actors in social networks and the situation that papers published in a journal cites other papers published in preceding journals. That is, journal papers and citation relationships from one paper to another correspond to nodes and directed edges from one node to another in a graph, respectively. In the case of the Web, Web pages and hyperlinks between pages similarly correspond to nodes and directed edges from one node to another, respectively.

Let us consider an adjacency matrix whose elements represent connections of two nodes. That is, the matrix E with elements E_{ij} is described as follows:

- (E_{ij})

E_{ij} of the adjacency matrix E is 1 if there is a directed edge from the node i to j, otherwise is 0. In this book, a matrix where all elements have nonnegative values is called nonnegative matrix. A matrix where all elements are greater than zero is called positive matrix. Only such matrices are considered in this section. Please note that both a nonnegative definite matrix and a positive definite matrix are different concepts. The adjacency matrix representing citation relationships (see Fig. 10.4a) is illustrated in Fig. 10.4b.

The prestige of the node i is denoted by p_i. Then the prestige p_i can be considered as summation of the prestige of all the nodes which have directed edges to the node i. Furthermore, a column vector p which have p_i as its components is considered. Then, using the transposed matrix E^T of the adjacency matrix E, the formula for calculating a new p, denoted by p', is described as follows:

- $p' \leftarrow E^T p$

Let an initial vector $p = (1, 1, --, 1)$. If the above formula is repeatedly applied, normalizing p so that $\Sigma p_i = 1$, a stationary solution for p will be obtained. Please note that there is another normalization method of carrying out division of p by the element with the largest absolute value. The way to

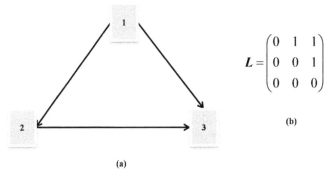

$$L = \begin{pmatrix} 0 & 1 & 1 \\ 0 & 0 & 1 \\ 0 & 0 & 0 \end{pmatrix}$$

(b)

(a)

Figure 10.4 (a) citation relationships, (b) adjacency matrix.

find the stationary solution is called the power method [Anton et al. 2002]. Premises for obtaining stationary solutions will be explained later.

The eigenvalue with a largest absolute value is called largest eigenvalue among all the eigenvalues of a matrix. In that case, p becomes an eigenvector corresponding to the largest eigenvalue (in this case, that absolute value is 1) of the matrix E^T. This is denoted by the following formula:

- $p = E^T p$

10.2.5 PageRank

Web pages also can be modeled with directed graphs as well as papers and books. A directed graph with weights may be used if needed. Here, PageRank of Larry Page and Sergey Brin [Liu 2007] will be explained. A PageRank of a certain page is decided by PageRanks of all the pages pointing to it. Let p_i be a PageRank of the page i. Moreover, let N_i be the number of links going out from the page i. Then, PageRank p_j of the page j is the sum of PageRank p_i of all the pages i coming into the page j with a weight $1/N_i$.

This scheme will be represented by using matrices. Let E be an adjacency matrix whose element E_{ij} is 1 if there is a link from the node i to j, otherwise 0. Furthermore, the following adjacency matrix L is defined using the elements of E. Both the matrices E and L are square matrices.

- $\left(L_{ij} \right) = \left(\dfrac{E_{ij}}{N_i} \right)$

Moreover, let p be a vector which has p_i as the i-th element. If the above matrix L is used, PageRanks will be calculated by the following formula:

- $p \leftarrow L^T p$

The method for computing PageRanks is fundamentally the same as the method for computing prestige. This is not surprising because Page and others referred to bibliometrix as a preceding work of PageRank. Generally, these schemes lead to solving the eigenvalue problems ([Watkins 2002], etc.) for the following square matrix M.

- $Mv = \lambda v$

Let λ and v be an eigenvalue and an eigenvector corresponding to λ, respectively. Supposing that this matrix satisfies certain conditions, the eigenvector v_t corresponding to the largest eigenvalue (let $\lambda = 1$) of a matrix will generally be obtained by the following algorithm called power method [Anton et al. 2002]. Here let v_0 be an initial vector for v_t. In each repetition, v_t is divided by the 1-norm $\|v_t\|$ for normalization. ε is the threshold value fixed beforehand.

(Algorithm) power method
1. $t \leftarrow 1$;
2. repeat {
3. $\quad v_t \leftarrow Mv_{t-1}$;
4. $\quad v_t \leftarrow v_t / \|v_t\|_1$;
5. $\quad t \leftarrow t + 1$;
6. } until $(\|v_t - v_{t-1}\|_1 < \varepsilon)$

It will be briefly shown that the eigenvector corresponding to the largest eigenvalue can be obtained by the power method. Assume that M has linearly independent eigenvectors $v_1, v_2, \ldots, v_n$. Furthermore, assume that there exists only one largest eigenvalue among eigenvalues, λ corresponding to the eigenvector v_i are ordered as follows.

- $|\lambda_1| > |\lambda_2| >= \ldots >= |\lambda_n|$

Let λ_1 and v_1 be the largest eigenvalue and the eigenvector corresponding to the largest eigenvalue.

If $Mv = \lambda v$ is substituted for the leftmost hand of the following expression, then the rightmost hand will be obtained.

- $M^2v = M(Mv) = M(\lambda v) = \lambda^2 v$

If this substitution is repeated, generally the following formula will be obtained:

- $M^i v = \lambda^i v$

By the way, the initial vector v_0 can be expressed as follows:

- $v_0 = c_1 v_1 + c_2 v_2 + \ldots + c_n v_n$

Here assume that c_i are real numbers and that c_1 is not equal to 0. That is, it is assumed that v_0 and v_1 are not orthogonal to each other. Then, we can obtain the following:

- $M^i v_0 = \lambda v_0 = \lambda_1^i (c_1 v_1 + c_2(\lambda_2/\lambda_1)^i v_2 + \dots + c_n(\lambda_n/\lambda_1)^i v_n)$

Here if clause $M^i v_0 / \lambda_1^i$ is considered, it will converge to $c_1 v_1$ by the assumption about the absolute value of the eigenvalue as i approaches infinity. It is normalized by dividing it by the norm of the vector or the component of the vector which has the largest absolute value if needed. Therefore, it has been shown that the eigenvector corresponding to the largest eigenvalue can be found by the power method.

Let us consider applying this method to the calculation of PageRank. In this case, verifying the justification of this algorithm leads to investigating whether the graph consisting of Web pages satisfies the conditions on which an eigenvector can be found by the above described power method.

It is known that if a graph generally denoted by a positive square matrix M is strongly-connected and aperiodic, the matrix M has only one largest eigenvalue according to the above formula and the vector in the power method converges to the eigenvector corresponding to the largest eigenvalue (Perron Frobenius theorem [Knop 2008]). Here the concepts of a strongly-connected component and cycle of a graph are defined as follows:

(Definition) Strongly-connected component of a graph

A strongly-connected graph is a directed graph whose arbitrary two nodes have bidirectional paths. A maximal strongly-connected subgraph of a directed graph is called strongly-connected component of the original graph.

(Definition) Aperiodic graph

If a node of a graph has a cycle of length one, or if the greatest common denominator of the lengths of all the closed circuits containing the node is one, the node is aperiodic. If a node has a cycle of longer than one, the node is periodic. Furthermore, if all the nodes of a graph are aperiodic, the graph as a whole is aperiodic.

After all it is necessary to make sure that the Web graph under consideration is both strongly-connected and aperiodic. From such a viewpoint, the justification of the above described algorithm will be considered by setting M to L^T. Clearly, however, the whole Web graph is not strongly-connected. Therefore, there is no guarantee that PageRanks can be always calculated. Then, what should be done? One answer to the problem will be given as follows.

Here a view of PageRank will be changed. First it is assumed that the user of the Web walks around pages at random according to hyperlinks. Each element of p represents a probability that the user stays in the page corresponding to the element. It can be considered that L^T is a transition matrix. This model about users' behaviors is called simple surfer model. However, as mentioned above, there is still no guarantee that an eigenvector can be obtained by the power method.

Now this simple surfer model will be extended by introducing a certain probability d (between 0.1 and 0.2) which is determined experientially. Assume that the user of the Web takes either of the following actions:

1. The user at random jumps to any page on the Web from the current page with a probability of d.
2. The user at random visits any page at the destination of a hyperlink going out of the current page with a probability of $(1–d)$.

In that the user can at random jump to arbitrary pages on the Web, this modified model is called extended surfer model. The probability that the user will stay at each page on the Web is calculated by the following formula:

- $$p \leftarrow d\left(\frac{1}{N}\right)p + (1-d)L^T p = d\frac{1}{N}(1, 1, ..., 1)^T + (1-d)L^T p$$

Here let N be the total number of the pages on the Web.

After all the matrix M in an eigenvalue problem is defined as follows:

- $$M = d\left(\frac{1}{N}\right) + (1-d)L^T$$

Here the sum of all the elements of any column of the matrix M must be one because they represent probabilities. Therefore, if a row in the matrix L has all zero entries, the node corresponding to the row is called dangling node. All the elements of such a dangling node are set to $1/N$. This adjustment is called stochasticity adjustment.

The adjacency matrix corresponding to the graph of pages illustrated in Fig. 10.5a is shown in Fig. 10.5b. Further the formula of PageRank for the matrix is shown in Fig. 10.5c. The graph denoted by the new transition matrix M constructed this way, is clearly a positive square matrix and satisfies the conditions that the graph is strongly connected and is aperiodic as well.

First, since every page on the graph can be considered as having a link directly coming from every page by action (1), the graph becomes strongly connected. Furthermore, in the strongly-connected graph, the cycle of every node is the same and is equal to the cycle of the whole graph itself. Since every node has also a "self-loop" link, the cycle as the whole graph is one as a result.

Thus, such an eigenvector, i.e., PageRanks, can be obtained by the power method. Here let the initial vector p_0 be $T(1/N, 1/N, ..., 1/N)$. This vector and a positive vector, where all the elements are positive, do not

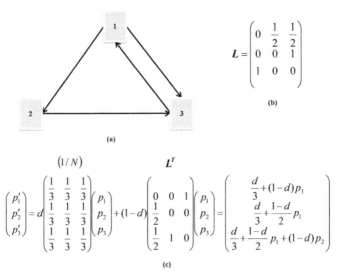

Figure 10.5 PageRank.

intersect orthogonally. Although PageRank is used for search engines (e.g., Google), it is not dependent on search results unlike HITS, which will be briefly described later. For the very reason, however, search engines based on PageRank may return results with less relevance to the search terms. On the contrary, due to the global characteristics of PageRank, PageRank is rather robust to the influences of spam actions such as increasing hyperlinks intentionally, as is one of the advantages of PageRank.

Next, the computational cost by the power method will be considered. In one repetition, the product of an $N \times N$ matrix and an N dimensional vector takes the cost of $O(N^2)$. Therefore, let R be the number of repetitions until convergence, the whole computational cost will be $O(N^2 \times R)$. Although the maximum number of repetitions cannot be presumed exactly, it has been reported that PageRank can converge for around 100 times in actual calculation. Moreover, if it is assumed that a Web graph is a sparse graph, data structures called adjacency lists can be used so as to reduce the previous cost of the product of the matrix and the vector to $O(N)$. In that case, the whole computational cost is $O(N \times R)$. If the total number N of the Web pages is set to 5×10^{10}, this improvement can be significant.

10.2.6 HITS

HITS (Hyperlink Induced Topic Search) [Liu 2007] determines ranks of Web pages like PageRank. The difference between HITS and PageRank will be described first. As already described, PageRank is calculated in advance for every page of the whole Web and is not dependent on individual queries.

On the other hand, HITS is calculated in principle every time the Web search system returns search results as a response to a specific query issued by the user and, of course, is dependent on the query. In reality, the search results are used after they are expanded as follows:

(Definition) A root page set, an extended page set, and a base page set
- Let r be a page included by a search result. The set of pages r is called root page set R.
- A set of pages which either come into or go out of r along a single path (i.e., the distance is one) is considered. This page set is called extended page set E.
- Furthermore, the root page set R and the extended page set E are merged and the resultant page set is called base page set B.

Next, the concepts of authority and hub about a page will be introduced. If a page is referenced (i.e., linked) by a lot of pages, it will be possible to say that such a page has a certain authority. The reference of a page by an important page adds greater authority to the referenced page than that by a less important page. The importance of a referencing page is measured by a value as a hub. A page referencing a lot of authoritative pages is a valuable hub. In short, the value as an authority of a page is determined by the aggregate values as hubs of pages referencing the page. Similarly, the value as a hub of a page is determined by the aggregate values as authorities of pages referenced by the page.

Then, let B be the base page and let a_i and h_i be the authority value and hub value of the page i, respectively. Moreover, by using the adjacency matrix E, the authority vector a and the hub vector h are represented as follows:

- $a = E^T h$
- $h = Ea$

If the power method is applied again, the algorithm for obtaining the two vectors is as follows:

(Algorithm) Calculate a hub and an authority of a page.
1. $t \leftarrow 1$;
2. repeat{
3. $a_t \leftarrow E^T h_{t-1}$;
4. $h_t \leftarrow Ea_{t-1}$;
5. $a_t \leftarrow a_t / \|a_t\|_1$;
6. $h_t \leftarrow h_t / \|h_t\|_1$;
7. $t \leftarrow t + 1$;
8. }until ($\|a_t - a_{t-1}\|_1 < \varepsilon$ and $\|h_t - h_{t-1}\|_1 < \varepsilon$)

Here let $a_0 = h_0 = (1/N, 1/N, \ldots, 1/N)^T$ and N be the total number of pages included by the base page set.

The adjacency matrix corresponding to an extended page set as shown in Fig. 10.6a is illustrated in Fig. 10.6b. The values of a_1 and h_1 based on the algorithm are illustrated in Fig. 10.6c. Now all that is needed is just to repeat this process.

- $a = E^T E a$
- $h = E E^T h$

These indicate that the authority vector a and the hub vector h are the eigenvectors of the matrices $E^T E$ and $E E^T$ (here $\lambda = 1$), respectively.

Here a few remarks will be made. There is no guarantee that there generally exists only one largest eigenvalue in the case of HITS unlike the case of PageRank. Therefore, the eigenvector corresponding to it is not necessarily unique. Moreover, what the eigenvector v converges to is dependent on initial values of v.

One of the features that HITS has is that HITS offers two different ranks, a hub and an authority, between which the user can choose. However, it is another feature of HITS that the result is greatly dependent on the search results of a search query. The reason for making an extended page set is to raise the recall ratio by covering as many possibly-relevant pages as possible. However, on the other hand, this will reduce the precision ratio.

Furthermore, once a portal site containing a lot of topics like Yahoo! is included in a base page set, HITS may rank pages about completely

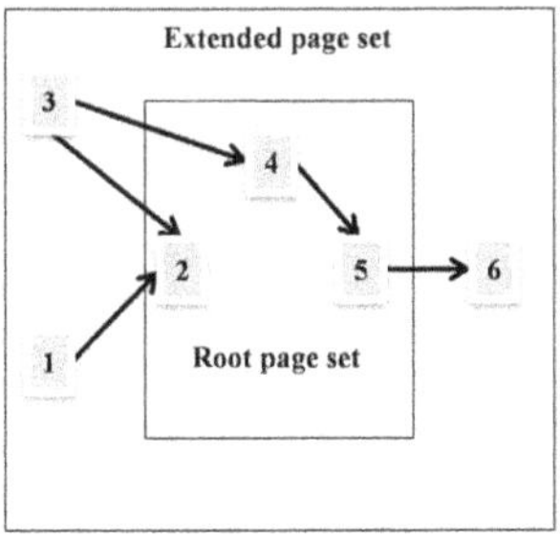

$$
E = \begin{pmatrix}
0 & 1 & 0 & 0 & 0 & 0 \\
0 & 0 & 0 & 0 & 0 & 0 \\
0 & 1 & 0 & 1 & 0 & 0 \\
0 & 0 & 0 & 0 & 1 & 0 \\
0 & 0 & 0 & 0 & 0 & 1 \\
0 & 0 & 0 & 0 & 0 & 0
\end{pmatrix}
\qquad
E^T = \begin{pmatrix}
0 & 0 & 0 & 0 & 0 & 0 \\
1 & 0 & 1 & 0 & 0 & 0 \\
0 & 0 & 0 & 0 & 0 & 0 \\
0 & 0 & 1 & 0 & 0 & 0 \\
0 & 0 & 0 & 1 & 0 & 0 \\
0 & 0 & 0 & 0 & 1 & 0
\end{pmatrix}
$$

(b)

$$a_0 = h_0 = (1/6 \quad 1/6 \quad 1/6 \quad 1/6 \quad 1/6 \quad 1/6)^T$$

$$
a_1 = \frac{E^T h_0}{\left\| E^T h_0 \right\|_1} = \begin{pmatrix} 0 \\ 2/5 \\ 0 \\ 1/5 \\ 1/5 \\ 1/5 \end{pmatrix}
\qquad
h_1 = \frac{E a_0}{\left\| E a_0 \right\|_1} = \begin{pmatrix} 1/5 \\ 0 \\ 2/5 \\ 1/5 \\ 1/5 \\ 0 \end{pmatrix}
$$

(c)

(a)

Figure 10.6 (a) (b) (c) HITS.

irrelevant topics contained by the portal page higher. Moreover, HITS is rather subject to the influences of SPAM actions such as attaching spurious links intentionally.

The computational cost of this whole algorithm is $O(N^2 \times$ number of repetition) like PageRank. It will be reduced to O ($N \times$ number of repetition) if the fact that Web is generally a sparse graph is utilized. Moreover, depending on the query, N in HITS is rather small in comparison with that in PageRank.

References

[Anton et al. 2002] Howard Anton and Robert Busby: Contemporary Linear Algebra. John Wiley & Sons (2002).

[ArnetMiner 2014] ArnetMiner http://arnetminer.org/ Accessed 2014

[Broder et al. 2000] Andrei Z. Broder, Ravi Kumar, Farzin Maghoul, Prabhakar Raghavan, Sridhar Rajagopalan, Raymie Stata, Andrew Tomkins and Janet L. Wiener: Graph structure in the Web. WWW9/Computer Networks 33(1-6): 309–320 (2000).

[CiNii 2014] CiNii http://ci.nii.ac.jp/ Accessed 2014

[CiteSeer^X 2014] CiteSeer^X http://citeseerx.ist.psu.edu/index Accessed 2014

[DBLP 2014] DBLP (Digital Bibliography & Library Project http://www.informatik.uni-trier.de/~ley/db/index.html Accessed 2014

[Google Scholar 2014] Google Scholar http://scholar.google.com/ Accessed 2014

[Hirsch 2005] J.E. Hirsch: An index to quantify an individual's scientific research output. In Proc. of the National Academy of Sciences 102(46): 16569–16572 (2005).

[Knop 2008] Larry E. Knop: Linear Algebra: A First Course with Applications (Textbooks in Mathematics), Chapman and Hall/CRC; 1 edition (2008).

[Liu 2007] Bing Liu: Web Data Mining–Exploring Hyperliks, Contents, and Usage Data. Springer (2007).

[Maslov et al. 2009] Sergei Maslov and S. Redner: Promise and Pitfalls of Extending Google's PageRank Algorithm to Citation Networks. arXiv:0901.2640v1 (2009).

[Watkins 2002] David Watkins: Fundamentals of matrix computations. John Wiley & Sons (2002).

[Web of Science 2014] Web of Science http://thomsonreuters.com/thomson-reuters-web-of-science/ Accessed 2014

Web Content Mining

This chapter explains search engines, information retrieval, page classification, page clustering, and microblog summarization as techniques for mining Web contents.

11.1 Search Engine

In order to search Web pages, the user usually consults Web search engines. Roughly speaking, tasks on the Web search engine side are divided into the following processes:

- Crawling Web pages
- Content analysis and link analysis of Web pages
- Indexing Web pages
- Ranking Web pages
- Search and query processing of Web pages

Before explaining these in detail, the flow of processes in a typical search engine will be reviewed briefly (see Fig. 11.1).

Generally, a search engine has some tasks to do before the user issues a query. First, the search engine collects pages from the whole Web and stores them in databases called a repository. This task is called crawling. More concretely, crawling programs follow links from seed pages as special origin pages and visit all the linked pages as destination pages and collect the contents of the pages. Crawling continues this process by setting the visited pages as new origin pages. Thus, crawling usually visits pages in a breadth-first manner and in a parallel fashion, starting with two or more seed pages. Crawling is performed periodically. Some search engines preferentially look for pages which are frequently updated or popular among the users.

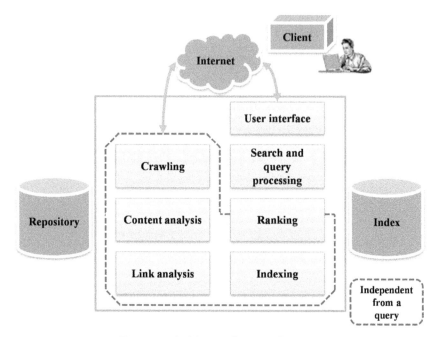

Figure 11.1 Architecture for a search engine.

Next, the search engine analyzes pages stored in the repository and extracts the URL and title of each page. Furthermore, words which can be used as search terms are extracted from the main part of the pages. This task is called content analysis. Both Web pages and search terms are managed with identifiers for distinction attached.

Then the search engine analyzes links of the pages and extracts anchor texts (i.e., texts on links) from them. This task is called link analysis. A link is stored as a pair of identifiers of an origin page and a destination page connected by the link together with the corresponding anchor text.

Then, the search engine prepares an index representing correspondence relationships between a page and all the search terms (more precisely, the identifiers) contained by the page, together with the positions of the terms within the page. The search engine creates another index (i.e., inverted index) representing correspondence relationships between a search term and all the pages containing the term, together with the positions of the term within the pages (i.e., the identifiers). With the aid of these indices, the search engine can find all pages containing specified search terms and all search terms contained by specific pages. The task that creates such indices is collectively called indexing.

Furthermore, based on the result of link analysis, some search engines adopt specific methods (e.g., PageRank of Google) to calculate ranks which express the importance of the crawled pages statistically (i.e., prior to search). Other search engines use similar methods (e.g., HITS) to calculate the ranks of pages dynamically, based not on the analysis prior to search but on the link analysis of search results. Generally, this task is called ranking.

Using the index described previously, the search engines collect pages containing a search term specified by the user at the time of searches. When two or more search terms are specified with spaces between them, some search engines (e.g., Google) calculate the set-product of the page sets obtained for each search term to obtain the final result. Other search engines calculate the set-union instead of the set-product. Furthermore, the search engines calculate the rank of each page belonging to a set of the obtained pages and synthesize it with the rank of the page calculated beforehand if necessary. The retrieved Web pages are sorted in descending order of ranks. A specified number (usually ten or twenty) of the pages (i.e., urls) are summarized into one SERP (Search Engine Result Page) with snippets generated for the pages, which is presented for the user to see one by one.

To calculate the final rank of the page on the spot, the search engines take into account not only how often the search term appears in the page (i.e., frequency of the search term) but also where the search term appears in the page (i.e., place for the search term). For example, titles and anchor texts are more important for the places than page bodies. This task is called query processing.

Among tasks for search engines, crawling, indexing, and ranking will be explained in detail below.

11.1.1 Crawling Web Pages

In a word, crawling collects a Web page and its url (uniform resource locator), a path to a host name and a file.

The basic algorithm of a crawler will be explained first in a very brief way (see Fig. 11.2).

1. Insert urls of Web sites as seeds into a data structure called pool.
2. Repeat the following steps until no more urls are found (i.e., the pool is empty) {
3.   Delete a url from the pool at its front end.
4.   Visit the page pointed by the url and collect urls contained by the page.
5.   Insert the newly collected urls into the pool at its rear end.
6.   Extract page information and link information from the accessed pages and store the former and the latter in the page repository and the link repository, respectively}.

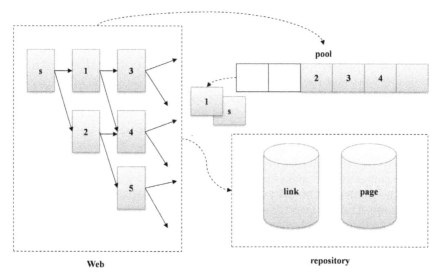

Figure 11.2 Crawler.

The program which performs such a series of processes is generally called Web crawler or Web spider, simply crawler or spider, respectively. In actual crawling, two or more crawlers perform distributed processing cooperatively. Moreover, if it is refused by the site administrator that the website and the pages are registered into the search engine by the robot, that is, if there exists a robots.txt file or robots meta-tag according to robots exclusion protocol [robotstxt 2014], then, needless to say, the crawler should respect them.

11.1.2 Indexing Web Pages

(1) Basic concept

The user utilizes Web search engines to find Web pages by specifying search terms. Then, the search engine must efficiently find Web pages containing the specified search terms. The mechanism for that purpose is indexing. The Web pages downloaded and stored in the page repository by the crawler are presented to the user at the time of search if demanded.

A search term is also called an index term because it is also used as a key for an index. First, the index terms which appear in Web pages are extracted. The basic unit is a word or term. If pages are written in English, words are usually delimited by spaces and the detection of words is rather straightforward. Then, how about Japanese? Since there are usually no spaces inserted between words or clauses in Japanese, detection of words is not so easy. Therefore, so-called morphological analysis is needed. In general, morphological analysis determines a part of speech and an inflection of a word using a dictionary. A lot of tools for morphological analysis called morphological analyzer are already available as free software such as Chasen for the Japanese language. Once the words are extracted from pages by morphological analysis, it is necessary just to implement the index for searching the documents by words as keys, using some dedicated access structures such as B+ trees or hash tables. Please note again that morphological analysis requires any word dictionary.

Methods without using morphological analysis include an N-gram index. Generally, N is the length of a character string (in Japanese) or the number of words (in English) extracted at once from a document or a page. If N is 1, 2, or 3, it is called unigram, bigram, or trigram, respectively.

For example, if this sentence is analyzed by bigram, it will become "For example", "example if" "if this", and so forth. The larger the number taken as N, the easier it is to find long terms. It is not necessary to prepare word dictionaries beforehand if an N-gram index is used based on characters. Let C be the number of distinctive characters, then the number of distinctive character strings indexed will become $O(C^N)$. In the case of Japanese, since C is on the order of 10^4, large numbers cannot be taken as N.

Please note that an N-gram index and a word-based index based on morphological analysis are not necessarily exclusive. That is, in creating an N-gram index for Japanese texts, it is also possible to take words obtained by morphological analysis as units instead of characters. For the sake of simplicity, a character-based N gram index will be explained below. However, if characters are replaced by words systematically, the outline still remains significantly the same.

The content analysis of Web pages and information retrieval techniques have much in common. The following subsection will explain information retrieval techniques.

(2) Structure of repository and index

A document identifier and a character string identifier are given to a page document (url) and a character string (i.e., N-gram), respectively. Such identifiers are determined so that they can be used as keys for hashing or sorting. First, a dictionary (that is, data structure for search) either with order or without order is built from a large collection of documents. Next, the position of a character string within a document, the type of the character string, and other information are attached to the character string. Types, such as url, title, anchor (i.e., text on a link), and meta-tag (description about a page or metadata), indicate places where the character string appears. The type of the character string is necessary because the importance of even the same character string varies according to contexts where it appears. Character case (i.e., upper or lower) and font size of the character string are also recorded as additional information. Moreover, if the character string has an anchor type, the document identifier of the origin page containing the anchor (i.e., anchor document identifier) is also recorded as additional information.

Here, only title and anchor are considered as types for the sake of simplification. Therefore, in the repository, a page has a title and a main part (body). The main part consists only of anchor texts and corresponding links (see Fig. 11.3a). A link consists of the origin page (i.e., document identifier), the destination page (i.e., document identifier), and the anchor text (see Fig. 11.3b). Furthermore, neither character case nor font size are considered.

The position within a document, the type, and the additional information of the character string are collectively called hit. Generally, according to the purposes of use, indices, not confined to N-gram indices, are classified into two types: a forward index and an inverted index.

They have the following separate roles.

- Forward index: Document identifier ->(character string identifier + hit) *
- Inverted index: Character string identifier ->(document identifier + hit) *

Generally, since two or more character strings (i.e., character string identifiers) may be contained by the document denoted by one document identifier and the same character string may appear two or more times, one or more "(character string identifier + hit)" are found in general. One or more "(document identifier + hit)" are similarly found for a certain character string identifier. "*" above expresses such situations. Please note that "(character string identifier + hit)" and "(document identifier + hit)" may be called posting.

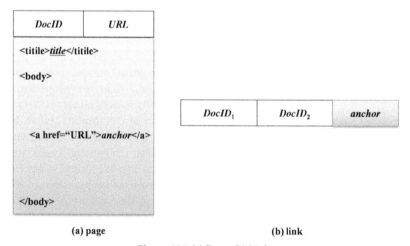

(a) page   (b) link

Figure 11.3 (a) Page, (b) Link.

In many cases, texts which appear in anchor links within a page describe pages which are referenced by the origin page. So anchor texts can also be considered as index terms of such pages. This technique is called anchor propagation, which is especially effective in indexing pages which contain no texts, such as images.

(3) Creation of an index

An algorithm for creating an index will be described. Basically, an N-length character string is extracted by shifting one character at a time and an index is created by using such strings. Please assume that there exist more than N characters in documents.

The algorithm takes N characters starting from the current position and makes a record which consists of a character string identifier, a document identifier, a position in the document, a type, and additional information. In case of anchor type, the record is copied and its document identifier is changed by that of the destination page of the anchor link so as to propagate anchor information to the page.

Let us create indices by using tables in RDBMS. For example, let a document identifier be a primary key in a B+ tree table, then the table can be used as a forward index of documents (see Fig. 11.4a). If another B+ tree index is created by using a character string identifier as a secondary key, it will be an inverted index of documents (see Fig. 11.4b). Please not that some applications need to compress not only the page data but also the index itself so as to reduce the storage size.

11.1.3 Ranking Web Pages

The ranking methods for Web pages can be classified as follows:

- (Static ranking) Calculate the ranks for all the crawled pages in advance of search.
- (Dynamic ranking) Calculate the ranks for all the searched pages by taking the similarity with search terms into consideration.

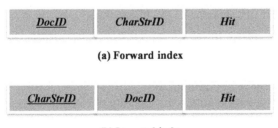

(a) Forward index

(b) Inverted index

Figure 11.4 (a) (b) Index.

For example, one of the representatives for static ranking is PageRank and representatives for dynamic ranking include HITS and the vector space model used in information retrieval in general. The vector space model is based on the similarity (e.g., cosine measure) between a vector of feature terms contained by a document and that of a search term specified by the user. In reality, the above two ranking methods are combined to determine the final rank of the retrieved page.

11.2 Information Retrieval Technique

In order to analyze the contents of pages which have been crawled, information retrieval techniques (IR) are used. Character strings (that is, equivalent to general textual documents) other than tags of pages are first analyzed in the following procedures.

- Morphological analysis: Segment texts into a sequence of words and determine parts of speech of the words. Segmentation is especially required of Japanese.
- Removal of unnecessary word: Remove unnecessary words (i.e., stop words) such as particles (in Japanese) and articles (in English) from a collection of words.
- Stemming: Take out a stem of a word to normalize it.

In this way, words which characterize textual documents are extracted. So such words are called feature terms of documents. Moreover, since they are used in the index for retrieval of documents, they are also called index terms.

11.2.1 Features

Next, weighting feature terms is considered.

Generally the weights d_{ij} of the feature term t_i contained by the document D_j are determined as follows:

- $d_{ij} = L_{ij}\, G_i / N_j$

Here each factor denotes one of the following:

- L_{ij}: A local weight. It is based on frequency of the feature term t_i in the document D_j.
- G_i: A global weight. It is based on distribution of the feature term t_i in the whole set of documents.
- N_j: A normalization factor. It is usually the length of the document D_j.

Furthermore, the concepts *TF*, *DF*, and *IDF* will be introduced.

- *TF*: A term frequency. It is the frequency of the feature term t_i in the document D_j.

- *DF*: A document frequency. It is the number of documents containing the feature term t_i divided by the number of all documents within the whole set.
- *IDF*: An inverse document frequency. It corresponds to the reciprocal of *DF*. However, it may not necessarily be an exact reciprocal. For example, logarithm of *DF* plus one is used as *IDF*.

Let us assume that no normalization is considered (i.e., always $N_j = 1$) and let *TF* and *IDF* be L_{ij} and G_i, respectively. Then the weight of the feature term in this case is equal to product of *TF* and *IDF* as follows:

- $d_{ij} = TF \times IDF$

Then, the weight defined in this way is often called simply *TFIDF*. There are variations for each factor such as L_{ij}. Please refer to text mining textbooks for detailed descriptions about the variations.

11.2.2 Vector Space Model

Next, the query processing based on the vector space model will be explained. The following matrix *D* is considered first.

- $[d_{ij}]$

D is called feature term-document matrix. Each column c_j of *D* is called document vector *d* because it expresses information about one document. Each row r_i of *D* is similarly called feature term vector because it expresses information concerning one feature term.

For example, the feature term document matrix as shown in Fig. 11.5 is considered. That is, a feature term Champagne is contained by both the document 1 and the document 3. The document 4 contains feature terms brandy and whisky.

On the other hand, a query can be viewed as a virtual document containing only search terms. So it is also expressed by the following column

	Doc1	*Doc2*	*Doc3*	*Doc4*	*Doc5*	*Doc6*
England	0	0	0	1	0	1
Whisky	0	0	0	1	1	0
Champagne	1	1	1	0	0	0
Sparkling wine	0	1	0	0	0	0
France	1	0	1	1	0	1
Brandy	1	0	0	1	1	0

Figure 11.5 Feature term-document matrix.

vector q, each element of which is the weight q_i (i.e., *TFIDF*) of the feature term t_i appearing in the query.

- $(q_1, q_2, \ldots, q_m)^T$

The result of a query is a set of documents similar to the query. The similarity of a query with documents is very important. For example, it is denoted by the cosine measure using the inner product of the document vector and query vector.

- Inner product: $d_j \cdot q = \sum_{i=1}^m d_{ij} q_i$

- Cosine measure: $\dfrac{d_j \cdot q}{|d_j||q|}$

Similarities calculated in this way are set to ranks of pages. Moreover, if any rank is available for each page in the result whether it is static or dynamic, such a rank and similarity are combined to produce a final rank. Documents are sorted in descending order of rank and are presented to the user as a result.

11.2.3 Accuracy of a Query Result

Next, the concept of relevance of a document to a query will be introduced. Relevance information determines a correct answer to a query as a subset of documents within the whole set. The relevance of each document to a query is usually determined by manual methods. That is, if a query is given, a set of correct documents will be obtained using the relevance information.

The performance of an information retrieval system can be evaluated by using the following two measures.

(Definition) recall

- $\text{recall} = \dfrac{|\text{a set of correct documents within the result}|}{|\text{a set of correct documents within the whole set}|}$

(Definition) precision

- $\text{precision} = \dfrac{|\text{a set of correct documents within the result}|}{|\text{a set of documents within the result}|}$

Figure 11.6 shows an example of relationships of a correct answer set and a set of retrieved results of a query. Generally there is a trade-off between recall and precision. Other measures of taking these two measures into consideration include the following *F* measure.

(Definition) *F* measure

- $F\ measure = \dfrac{2}{\dfrac{1}{recall} + \dfrac{1}{precision}}$

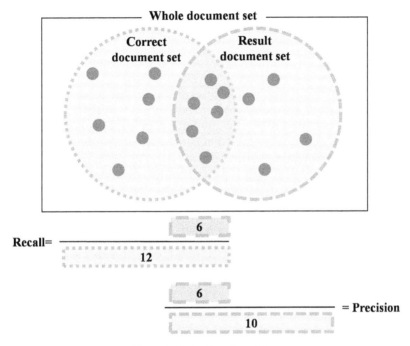

Figure 11.6 Recall and precision.

11.2.4 Miscellaneous Topics

Some topics will be explained in relation to information retrieval techniques.

(a) Relevance feedback

Information retrieval is essentially interactive. It is possible to modify the query itself so as to better fit the result to the user's demand. This is called relevance feedback ([Liu 2007] etc.). Thus the user divides a set of documents as a result returned by the information retrieval system into two groups: relevant documents (R) and irrelevant documents (IR). Then the system modifies the query (q) by using the average of feature vectors of each group to generate a new query (q'). After that, the system executes the new query so as to return a new result. The user repeats this process until the satisfactory result is obtained.

The algorithm of Rocchio, a typical method, is shown below.

(Algorithm) The algorithm of Rocchio

- $q' = \alpha q + \dfrac{\beta}{|R|}\sum_{d \in R} d - \dfrac{\gamma}{|IR|}\sum_{d \in IR} d$

Here α, β, and γ are positive constants and are determined with heuristic methods. It is expected that relevance feedback improves recall and precision.

(b) Signature

In information retrieval, signature [Han et al. 2001] is also used. For example, a signature of a word is expressed by a bit string as a result of hashing it. On the other hand, a signature of a document is expressed by the logical sum of the signatures of words contained in the document. A query is expressed by the logical sum of the signatures of search terms contained in the query like normal documents. If logical product of the signature of a query and that of a document is equal to that of the original query, then the document is a candidate of result documents.

(c) Jaccard coefficient and Tanimoto coefficient

As similarity used for searching pages, not only the cosine measure of feature terms based on *TFIDF* but the following coefficients are also utilized.

(Definition) Jaccard coefficient
- Suppose that a shingle is q-gram (continuous q length of tokens).
- If a set of shingles contained in the document d is denoted by $S(d)$, the Jaccard coefficient can be defined as follows:
- Jaccard coefficient $(d_1, d_2) = \dfrac{|S(d_1) \cap S(d_2)|}{|S(d_1) \cup S(d_2)|}$

If a general set is considered instead of $S(d)$ here, the Jaccard coefficient will define the similarity of two sets. Moreover, in information retrieval, Tanimoto coefficient may be used as similarity between documents.

(Definition) Tanimoto coefficient
Let the feature vector of the document d_i be $d_{i'}$, then the Tanimoto coefficient is calculated as follows.
- Jaccard coefficient $(d_1, d_2) = \dfrac{|S(d_1) \cap S(d_2)|}{|S(d_1) \cup S(d_2)|}$

In a special case where each element of a feature vector is dichotomous (i.e., 0 or 1), a Tanimoto coefficient becomes equivalent to a Jaccard coefficient about the sets represented by the feature vector since each element of such a vector indicates either the existence or non-existence of the corresponding element of the set.

In addition, the links of pages can also be simultaneously used for calculating the similarity for search of pages. For example, such methods include HITS-based methods and cocitation-based methods, which will be described later.

(d) LSI

Next, LSI (Latent Semantic Indexing) [Liu 2007] will be explained. LSI is executed by the following steps

1. Perform singular value decomposition of feature term-document matrix $A_{t \times d}$ [Anton et al. 2002] and obtain the product of matrices $U_{t \times r} S_r V_{r \times d} T$ as a result.
   Here $U_{t \times r}$ and $V_{r \times d}$ are orthogonal matrices. S_r is a diagonal matrix with the rank r. Diagonal elements of S_r are singular values, $\sigma >= \ldots >= \sigma_r > 0$.
2. Replace each original document vector by the corresponding k-dimension document vector, using the matrix S_k which is made by selecting the $k(<r)$ largest singular values.

By making the frequencies of some hidden feature terms larger, LSI enables the user to discover semantic similarity between documents or that between a document and a query which was latent based on the original feature terms. In other words, LSI can actualize conceptual structures, called k-concept space, hidden under the influence of synonymous terms. Thus, LSI cannot only remove some noises, but can also reduce the dimensions of document vectors. LSI can be used as preprocessing of clustering or visualization.

(e) Use of association rules

Document mining using association rules will be described. If each document is considered to be a transaction, feature terms contained by the document are considered to correspond to items within the transaction. Then mining association rules can be applied to documents. An association which frequently appears between consecutive terms within a document may constitute a compound term (i.e., phrase). If compound terms are correctly detected, automatic tagging of documents and deletion of meaningless results can be performed. If association analysis is applied to a set of search terms specified as a query, suggestion of frequently co-occurring search terms can be realized.

11.3 Classification of Web Pages

Search of Web pages has been explained so far in detail. Here, classification of Web pages will be described. As preparation, classification techniques for general documents will be described first. Yang and others in their paper [Yang et al. 1999] compared the following three techniques:

- Support vector machine
- k-nearest neighbor
- Naive Bayes

Although this paper compared linear least-squares method [Manning et al. 1999] and neural network [Mitchell 1997] as well, only the above three are described due to popularity as classification techniques.

11.3.1 Support Vector Machine

A support vector machine [Burges 1998] has been introduced by Vapnik. In space which can be linearly separated, the following hyperplane, called separating hyperplane, determines whether given data are positive or negative with respect to a certain class.

- $w \cdot x - b = 0$

Here a vector x denotes a document to be classified. w and b are learned from a training set which can be linearly separated.

Here $D = \{(y_i, x_i)\}$ is a training set. In case of $y_i = +1$ or -1, x_i is a positive example or a negative example of a certain class, respectively. The support vector machine determines w and b satisfying the following formulas so that 2-norm (i.e., Euclidean distance) of w will be the minimum.

- $w \cdot x - b \geq + 1 (y_i = +1)$
- $w \cdot x - b \leq - 1 (y_i = -1)$

A set of training data whose distance from the separating hyperplane is equal to $1/\|w\|$ are called support vectors (see Fig. 11.7). In fact, w and b are determined only by these support vectors and the other data are not required.

In general, of course, data space cannot always be linearly separated. Such data are mapped into linearly separable space of higher dimensions. Let φ be such a mapping, all one has to do is to replace x by $\varphi(x)$ in the upper formula.

11.3.2 k-nearest Neighbors

The method k-nearest neighbors, shortly called k-NN [Han et al. 2001], finds a subset of the k most similar documents with respect to the given document out of the given training set where the classes of documents are known in advance. Then the weights about the candidate classes of the document are determined by aggregating the similarities between the original document and the documents in k-NN. As similarity, for example, the cosine measure as to the feature term vectors of documents is used.

First, given document feature vectors, the following scores $y(x, c_j)$ for two or more candidate classes c_j are calculated.

- $y(x, c_j) = \sum_{d_i \in k-NN} sim(x, d_i) \, y(d_i, c_j) - b_j$

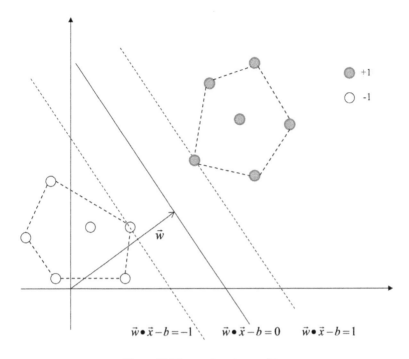

$$\vec{w} \bullet \vec{x} - b = -1 \qquad \vec{w} \bullet \vec{x} - b = 0 \qquad \vec{w} \bullet \vec{x} - b = 1$$

Figure 11.7 Support vector machine.

Here if d_i is classified into a class c, then $y(d_i, c)$ is 1, otherwise it is 0. $Sim(x, d_i)$ denotes the similarity between the documents x and d_i.

Then assignment of a document to a class is determined by using a threshold value for the score. That is, if the score exceeds the threshold value, the document belongs to the class, otherwise it does not belong to the class.

Some remarks will be made here. An optimal threshold value for b_j is learned by using a subset of training set. For example, b_j which maximizes F value is determined as optimal. If b_j is allowed to be nonoptimal, use of a constant as b_j makes prior learning needless.

By this method, one document is fundamentally allowed to belong to two or more classes. There may also be a variation to these. For example, only c_j which maximizes the score denoted by the above formula is chosen as the class. In that case, the value of $y(x,c)$ is set to 1, otherwise it is set to 0. In that case, if both the similarity sim and the threshold b_j are constant, this score equals just majority voting.

11.3.3 Naive Bayes

The naive point in classification of documents using Naive Bayes is the assumption that given a class, the conditional probabilities of one feature term and of another in the class are independent. For example, let us consider classification of documents by the following formula [Mitchell 1997]. Classification of documents *Doc* is equivalent to finding the class v_j which maximizes this value.

- $v = argmax_{v_j} \, P(v_j) \prod_{i \in Positions} P(a_i = W_k | v_j)$

Here *Positions* is a set of the positions of the terms in *Doc* and a_i is a term in *i*-th position of the document. Some variables are also introduced as follows:

- *Vocabulary*: The vocabulary in the training data D
- D_j: A set of documents among D which belong to v_j
- T_j: The single document which can be made by concatenating all the elements of D_j
- N: The total number of positions of terms in T_j
- N_k: The frequency of the term W_k in T_j

Moreover the following probabilities are learned in advance.

- $P(v_j) = \dfrac{|D_j|}{|D|}$
- $P(a_i = W_k | v_j) = P(W_k | v_j) = \dfrac{N_k + 1}{N + |Vocabulary|}$: The probability that the term extracted from the document belonging to the class v_j is W_k. It is independent of *i*. Please note that m-estimate is used so as to reduce the bias produced by the probability resulting to 0.

At least, according to Yang et al., when the performance of the above methods were compared, the methods were ranked in order of good performance.

- support vector machine > *k*-nearest neighbors >> Naive Bayes

11.4 Clustering of Web Pages

Next, clustering of Web pages will be described. As preparation, the clustering techniques for general documents will be described first. The comparative study [Steinbach et al. 2000] of such techniques compared the following three measures to determine two clusters to be merged in hierarchical agglomerative clustering.

- Intra-Cluster Similarity: Consider the total sum of the similarities of the cluster's centroid and the documents in the cluster. Let *Sim(X)* be such a similarity of the cluster X and let C_3 be a new cluster made by

merging two clusters C_1 and C_2. The clusters C_1 and C_2 are merged that maximize the value $\text{Sim}(C_3)\text{-Sim}(C_1)\text{-Sim}(C_2)$.
- Centroid Similarity: Two clusters are merged that maximize the similarity between the centroids.
- UPGMA (Unweighted Pair Group Method with Arithmetic Mean): Two clusters are merged that maximize the average value of the similarity of all the document pairs each of which is contained in a separate cluster.

Note that the cosine measure based on the feature terms was used for the similarity between documents or clusters.

Steinbach and others concluded that UPGMA was the best among the above three methods. Furthermore, they also compared the *k*-means, the bisecting *k*-means, and hierarchical agglomerative clustering using UPGMA and concluded that the bisecting *k*-means was the best among them. The algorithm of the bisecting *k*-means method is described below.

(Algorithm) The bisecting *k*-means method
1. Repeat the following procedure until the cluster number reaches *k* {
2. Choose one cluster according to a suitable measure;
3. Divide the cluster into two by using the *k*-means method ($k = 2$) and substitute the original cluster by the new two clusters;}

Here as candidate clusters for division, those with the largest size or the smallest intra-cluster similarity are chosen.

Next, the technique of clustering Web pages is described.

In addition to clustering mainly based on feature terms contained by Web pages, there are other methods based on the structures of Web pages. In order to cluster the results of Web search engines, some systems involve not only the feature terms within the pages but also the incoming and outgoing links of the pages in the similarity measure (for example, cosine measure) between pages.

For example, while Wang and others [Wang et al. 2002] used the *k*-means method with a weighted cosine measure involving the above three, Modha and others [Modha et al. 2000] used the *k*-means method extended with weighted inner products involving the above three aspects. These approaches commonly use the concept co-citation, which is defined as follows.

(Definition) co-citation
- If both the documents *A* and *B* are cited from *C*, it is said that *A* and *B* are co-cited.

The concept of co-citation is illustrated in Fig. 11.8.

As for the citation relationships illustrated in Fig. 11.8a, there exists a citation matrix L illustrated in Fig. 11.8b. The element c_{ij} of the symmetric

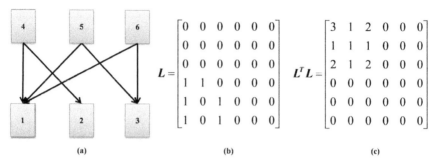

$$L = \begin{bmatrix} 0 & 0 & 0 & 0 & 0 & 0 \\ 0 & 0 & 0 & 0 & 0 & 0 \\ 0 & 0 & 0 & 0 & 0 & 0 \\ 1 & 1 & 0 & 0 & 0 & 0 \\ 1 & 0 & 1 & 0 & 0 & 0 \\ 1 & 0 & 1 & 0 & 0 & 0 \end{bmatrix} \quad L^T L = \begin{bmatrix} 3 & 1 & 2 & 0 & 0 & 0 \\ 1 & 1 & 1 & 0 & 0 & 0 \\ 2 & 1 & 2 & 0 & 0 & 0 \\ 0 & 0 & 0 & 0 & 0 & 0 \\ 0 & 0 & 0 & 0 & 0 & 0 \\ 0 & 0 & 0 & 0 & 0 & 0 \end{bmatrix}$$

(a)   (b)   (c)

Figure 11.8 Co-citation.

matrix $L^T L$ illustrated in Fig. 11.8c expresses the number of times when the document i and the document j are co-cited.

If A and B are co-cited, it is interpreted that they are semantically related. In the context of Web pages, co-cited documents correspond to pages which are simultaneously linked from one page.

The clustering algorithm based on co-citation used by the system of Pitkow et al. [Pitkow et al. 1997] will be described below.

(Algorithm) Clustering based on co-citation
1. For a set of documents with citation information added, the number of citations is counted for every document. Consider only documents with the number of citations equal to or larger than a certain threshold as the targets of further processing.
2. Make a pair of co-cited documents and count the number of citations. A list of such pairs is called pairlist.
3. One pair is chosen from the pairlist.
4. Look for other pairs containing at least one document of the pair in the pairlist. Repeat this step for found pairs until there is no such pair. Let a set of all documents (pairs of documents) obtained in this way be one cluster.
5. If there is no pairs in the pairlist, then terminate. Otherwise, go to Step 3.

In other words, the algorithm computes a transitive closure based on co-citation relationships. The transitive closures correspond to clusters. Moreover, only if the document i cites the document j in an adjacency matrix E described so far, then the element $(i, j) = 1$. The element (k, l) of the matrix $E^T E$ expresses a similar relationship between the documents k and l and the matrix can be a co-citation index.

11.5 Summarization of Microblogs

In this subsection, summarization of not general documents but microblog (e.g., twitter) articles will be described. Articles in twitter, called tweets, are produced more rapidly than those in general blogs. It is especially important to summarize the contents of a set of articles which are obtained by searching for a series of articles with emergent or interesting topics.

The following procedures summarize tweets (see Fig. 11.9).

(Algorithm) Summarization of tweets
1. Search for a set of articles relevant to a specified topic. For example, if a hash tag (i.e., a topic prefixed by #) is available, related articles can be acquired using streaming API with the hash tag specified as the search key.
2. A burst is detected from the article set in real time. For example, a burst is defined as follows.

(Definition) Burst
- A period during which the average of intervals between two succesive articles becomes extremely shorter than the average of intervals during the previous period is called burst.
3. An article set with the maximum degree of covering the article set restricted to the burst period is considered as a summary of the topic.

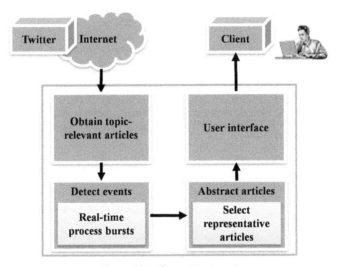

Figure 11.9 Abstracting tweets.

The degree of covering of a certain document d_1 is defined as the total sum of the similarity between d_2 and another document, which is defined as follows

(Definition) Similarity between documents

- similarity between documents $(d_1, d_2) = \dfrac{|S(d_1) \cap S(d_2) \cap I|}{|S(d_1) \cup I|}$

Here $S(d)$ is a set of feature terms contained by the document d; I is a set of important terms contained by a document set and is determined based on *TFIDF*.

References

[Anton et al. 2002] Howard Anton and Robert Busby: Contemporary Linear Algebra, John Wiley & Sons (2002).

[Burges 1998] Christopher J.C. Burges: A Tutorial on Support Vector Machines for Pattern Recognition. Data Mining and Knowledge Discovery 2(2): 121–167 (1998).

[Han et al. 2001] Jiawei Han and Micheline Kamber: Data Mining: Concepts and Techniques. Morgan Kaufmann (2001).

[Liu 2007] Bing Liu: Web Data Mining—Exploring Hyperlinks, Contents, and Usage Data. Springer (2007).

[Manning et al. 1999] Christopher D. Manning and Hinrich Schütze: Foundations of Statistical Natural Language Processing. The MIT Press (1999).

[Mitchell 1997] Tom M. Mitchell: Machine Learning. McGraw-Hill (1997).

[Modha et al. 2000] Dharmendra S. Modha and W. Scott Spangler: Clustering hypertext with applications to web search. Research Report of IBM Almaden Research Center (2000).

[Pitkow et al. 1997] James Pitkow and Peter Pirolli: Life, Death and lawfulness on the Electronic Frontier. In Proc. of ACM SIGCHI, pp. 383–390 (1997).

[robotstxt 2014] robots.txt http://www.robotstxt.org/ Accessed 2014

[Steinbach et al. 2000] Michael Steinbach, George Karypis and Vipin Kumar: A Comparison of Document Clustering Techniques. In Proc. of KDD Workshop on Text Mining (2000).

[Wang et al. 2002] Yitong Wang and Masaru Kitsuregawa: Evaluating Contents-Link Coupled Web Page Clustering for Web Search Results. In Proc. of the eleventh international conf. on Information and knowledge management, pp. 499–506 (2002).

[Yang et al. 1999] Yiming Yang and Xin Liu: A re-examination of text categorization methods. In Proc. of 22nd ACM International Conf. on Research and Development in Information Retrieval, pp. 42–49 (1999).

Web Access Log Mining, Information Extraction, and Deep Web Mining

First in this chapter the basic techniques for Web access log mining and the applications including recommendation, site design improvement, collarborative filtering, and Web personalization will be described. Next, the techniques for extracting information from the generic Web and mining the deep Web including social data, will be explained.

12.1 Web Access Log Mining

12.1.1 Access Log Mining and Recommendation

Web access log mining is to analyze the access histories of the users who accessed a website [Liu 2007]. The results of the analysis are mainly used for recommendation of the pages to other users or to re-design the website. When human users and so-called Web robots access Web sites, an entry including the IP address, the accessed time, the demanded page, the browser name (i.e., agent), the page visited just before visiting the current page, and search terms is recorded in the Web access log (see Fig. 12.1) [Ishikawa et al. 2003].

After data cleansing such as removal of unnecessary Web robots' histories is performed for the access log, sessions are extracted from the log. Then user models are created by classifying or clustering users.

Basically, whether the visitor is a human or a Web robot can be known by the records as to whether the visitor accessed the file called robots.txt. This is because the Web robots are recommended to follow the policy of the Web site about acceptance of robots (i.e., robot exclusion agreement)

①**133.86.XX.XXX** ②- ③- ④[2006-04-01 10:27:07 +0900]
⑤**"GET /index.html HTTP/1.1"** ⑥**200**⑦**9554**
⑧**"http://www.tmu.ac.jp/academics.html "**
⑨**"Mozilla/5.0 (compatible; MSIE 9.0; Windows NT 6.1; Trident/5.0)**
"

①**hostname** ②ident ③authuser ④date
⑤**request** ⑥status ⑦bytes
⑧**refer**
⑨**useragent**

Figure 12.1 An example of web access log data.

described by the file. Moreover, the distinction between human visitors and
Web robots can also be made by checking whether the visitors are included
in the robot list which has been created in advance. However, these methods
are not effective against malicious Web robots or new Web robots which
have not been registered yet. In that case, it is necessary to detect such Web
robots by their access patterns and this task itself will be a kind of Web
access log mining [Tan et al. 2002]. Anyhow, assume for simplicity that the
Web access log is cleansed, that is, Web robot accesses are removed from
the access log by a certain method.

A sequence of page accesses by the same user is called session. Basically
whether the visitor is the same user or not is judged by its IP address.
Generally, however, there is no guarantee that the same IP address (for
example, dynamic IP address) represents the same user. Therefore, in order
to correctly identify the same user, it may be necessary to combine other
information (for example, agent).

It is usually assumed that the time interval between one access and the
subsequent access within a session is less than 30 minutes. Thus, a session
is united. As for other methods of extracting a session from the access log,
it is possible to use a threshold value for the duration of a whole session
or to include page accesses in a session under consideration which contain
its preceding page access.

How to extract access patterns based on transition probability will be
explained below.

The transition probability P from the page A to the page B (denoted by
$A \Rightarrow B$) can be calculated as follows (see Fig. 12.2).

- $P(A \Rightarrow B) = \{$the number of transitions from A to $B\}/\{$the total number
  of transitions from $A\}$

Furthermore, the transition probability P of the path of pages ($A \Rightarrow B
\Rightarrow C$) is calculated as follows.

- $P(A \Rightarrow B \Rightarrow C) = P(A \Rightarrow B) \times P(B \Rightarrow C)$

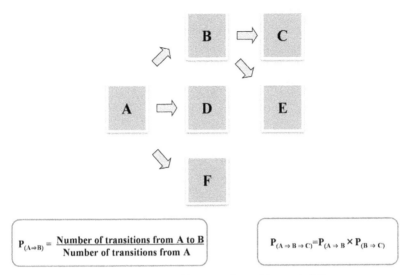

Figure 12.2 User access history and transition probability.

Here the above method for probability calculation assumes that each transition is an independent event.

The user model is constructed by grouping (i.e., clustering) similar users based on the access patterns extracted in this way. Below, by taking recommendation as an example, applications of the mining result (namely, rules) will be explained. Basically, a rule whose transition probability is equal to or higher than the threshold value is applied to the user who has accessed a certain page. That is, similarities between users and the user models are calculated by a certain method and then the most plausible user model is determined. Then, based on the page where the user currently stays, another page is recommended to the user according to access patterns most probable in the selected model.

For example, the following can be considered as recommendation schemes only based on transition probability (see Fig. 12.3).

- Path recommendation: Analyze a path (i.e., sequence) of pages which a user frequently accesses and recommend the whole path at once.
- Recommendation based on link prediction: Recommend only the last page of a path which users frequently access.
- Recommendation based on access history: Recommend a page based on the pages accessed so far as well as the current page. Please note that, unlike the above described probability $P(A \Rightarrow B \Rightarrow C)$, the transition probability $P_H(A \Rightarrow B \Rightarrow C)$ of path $A \Rightarrow B \Rightarrow C$ in the recommendation based on access history is calculated by the following formula:
$P_H(A \Rightarrow B \Rightarrow C) = \{$The number of transitions from A to C via $B\}/\{$the total number of transitions from A via $B\}$

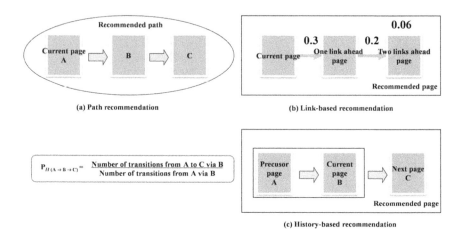

(a) Path recommendation

(b) Link-based recommendation

(c) History-based recommendation

Figure 12.3 Transition probability-based recommendation.

Therefore, if the user comes to B from pages other than A (i.e., A'), this method uses the probability of $P_H (A' \Rightarrow B \Rightarrow C)$ instead. Generally the Apriori algorithm and its variations for mining association rules are applicable to increasing the efficiency of calculation of this probability P_H.

Moreover, recommendation methods which give weight to transition probabilities by considering other information such as structures of Web pages include the following (see Fig. 12.4).

- Recommendation weighted by links: If a page is linked by N pages, simple methods use N as a weight of the page. More sophisticated methods apply techniques of Web structure mining (e.g., PageRank) to pages within a website and determine the weights of the pages based on the ranks of the pages.
- Recommendation weighted by visits of users: A higher weight is given to pages which are characterized by visits of users, such as pages where users stayed for a long time or pages which users frequently visited in the nearest past.

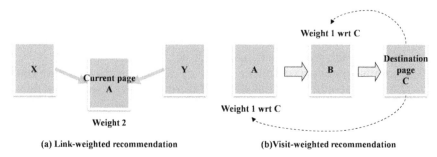

(a) Link-weighted recommendation

(b) Visit-weighted recommendation

Figure 12.4 Weighted transition probability-based recommendation.

The authors [Ishikawa et al. 2003] have considered reduction in click-through counts and users' time to destination pages as measures for the effectiveness of recommendation. Based on these measures, the authors have discovered that link-weighted recommendation is at least more effective than other methods with no weights.

12.1.2 Clustering Access Patterns

Here, let us consider clustering access patterns. Each session is represented as a sequence of page identifiers. Then, sessions are clustered based on the similarity of the sequences. As a research which clusters user access patterns, the authors' work [Ishikawa et al. 2003] will be explained.

This research has been aimed at mining the access history for one Sunday, of a movie theater site, which actually exists in Japan. This work uses bigram expressions of transition. For example, a transition from the page A to the page B is denoted by $A \Rightarrow B$.

In general, if there are N pages in all in a website, $N \times (N-1)$ distinctive transitions are possible. For the sake of simplicity, each session is represented by the feature vector whose element indicates existence of a transition during a session, represented by the bigram expression. Instead of the feature vector using bigram expressions of page transitions, the representation of a session by a sequence of page transitions [Fu et al. 1999] may be used, considering the length of stay at each page.

The applicable clustering techniques include k-means [Shahabi et al. 1997], BIRCH [Fu et al. 1999], and other methods [Yan et al. 1996]. In the authors' experiments, hierarchical agglomerative clustering was applied to clustering the sessions in the Web access log together with the Ward method based on the Euclidean distances as dissimilarity. In the experiments, as the shortest distance between similar clusters changed drastically when the number of clusters decreased from 6 to 5, six target clusters were created. As a result, the page transition *NextRoadshow* $\Rightarrow$ *Roadshow* was the most frequent in the 6th cluster. Incidentally, the tool visualizing a dendrogram intelligibly was developed and used for detection of changes of the distance between similar clusters.

The authors have interpreted that the user usually expects that the Roadshow page has a schedule for this week and the *NextRoadshow* page has a schedule for the next week. According to the access log on Sunday, the user accessed the page *NextRoadshow* in order to see the schedule for the next week.

On the other hand, the roadshow schedule managed a week from Saturday of this week to Friday of the next week as a unit at this website. The roadshow schedule for the next week, contained by the page *NextRoadshow*, was moved (not copied) from the page *NextRoadshow* to the page Roadshow on Saturday, i.e., the day preceding the day when the user actually accessed the site. Furthermore, between Saturday of this week and Tuesday of the next week, the page *NextRoadshow* was empty for update (!). Then, the user noticed the visit of the "wrong" page (*NextRoadshow*) during access of this day (Sunday) and moved to the "right" page (Roadshow).

It turned out, as a result of mining, that such user accesses had occurred frequently. This indicates user actions contrary to the site administrator's expectation. That is, the parts enclosed with the parenthesis ("") in the above paragraph differ in recognitions of the users and of the site administrator. In other words, the analysis result can suggest that the site design was not appropriate in this case. A situation like this could be overlooked during the stage of site design in many cases. Although the result can also be harnessed in recommendation of pages, it can be more effective in advising the site administrator that the inappropriate site design should be improved.

12.1.3 Collaborative Filtering and Web Personalization

Generally, the following applications for websites that use recommendation techniques based on access log mining can be considered.

1. Recommend a relevant page or a set of relevant pages without changing the website.
2. Change the existing pages dynamically and recommend the new ones.
3. Propose to the website administrator the redesign of the website, i.e., to permanently change the current pages.

The applications (1) and (2) correspond to usual Web recommendations. Especially, (2) is called Web personalization [Mobasher et al. 2000]. The application (3) corresponds to the example explained in the preceding section. Anyway, the result of Web access log mining can help the user to efficiently reach the target pages. In addition, some recommendation systems explicitly obtain information about the tastes of the users by questionnaires in advance in order to build the user models while other recommendation systems analyze the user's history data on browsing pages and buying products.

The following examples are among the representative cases of recommendation performed by a lot of commercial websites of the present.

*   Amazon.com recommends to the customer books which were bought by other customers who bought a book bought by the former customer. This case uses co-reference and co-citation for identification of user models and recommendation of items, respectively.
*   Yahoo!auction allows parties involved with dealings to evaluate each other. The evaluation is open to the public and used by a third party to conduct new dealings.

Thus, recommendation systems recommend products according to the user's tastes, products in the categories relevant to products bought by the user, or products bought by those who bought products bought by the user. Recommendation based on action patterns of similar users like the last example (used by Amazon) is called collaborative filtering. The system architecture of general collaborative filtering is illustrated in Fig. 12.5. Needless to say, technologies of data mining, such as clustering, classification, and association analysis (also, series data mining), which have been described so far in this book, are applicable to these recommendation systems.

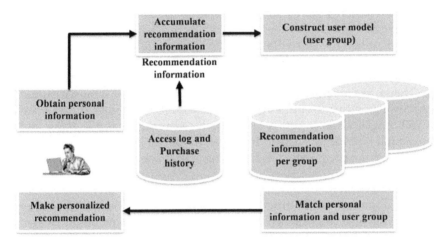

Figure 12.5 Architecture for collaborative filtering.

12.2 Information Extraction

12.2.1 Tasks in Information Extraction

Let us assume that there is a project to create meta-search engines. In order to reuse the results of existing search engines for that purpose, it is necessary to extract information about each page (e.g., url) from SERP (Search Engine Result Pages), which usually list ten or twenty retrieved pages. Similarly, in order to use the search results of Internet shopping sites for constructing meta-shopping search engines or comparison shopping sites, it is necessary to correctly extract each item from the result pages of individual search engines or comparison sites.

Moreover, in order to analyze a researcher's activities from the authored papers or the affiliated academic societies, it is necessary to correctly extract bibliometric information about the papers from digital libraries as well as information about the committee members from the sites of international conferences and the editorial members from those of academic journals.

Furthermore, in a geographic information system (GIS), information about landmarks, such as positions and explanations of buildings or facilities, are often needed. In such cases necessary information can be extracted from relevant pages such as Wikipedia.

Thus, in order to build these applications, it is necessary to identify such entities and extract values of their attributes out of Web pages as sources of information. The contents of Web pages used as an input, especially, portions except links are usually unstructured (i.e., flat) texts or semi-structured texts (HTML). The semi-structured texts include tables and lists. On the other hand, data which serve as an output are fundamentally structured data, which can be expressed by the tuples (i.e., records) of relational databases. However, missing values, collections of values as well as erroneous values may be included in the values of attributes. Therefore, in general, output data can be considered to be semi-structured data (for example, XML).

Therefore, it is necessary to extract structured data or semi-structured data corresponding to entities from flat or semi-structured texts contained in Web pages as an input. A task which performs such discovery and conversion of entities is called information extraction. The program that performs this task is called information extraction wrapper, or shortly, wrapper.

Semi-structured texts as input data of information extraction are either dynamically created from structured databases (e.g., relational databases) by servers when they are searched from the deep Web sites or manually created as rather static pages beforehand like Wikipedia articles. Furthermore, their variations can be considered for each category. Here, the information extraction from static pages will be explained. Methods targeted at dynamic pages will be treated in the section of mining the deep Web.

12.2.2 Issues in Information Extraction

There are technical issues related to information extraction as follows.

Since entity information is usually extracted from two or more sources of information (i.e., websites), the wrappers need to handle two or more kinds of data structures as the input. In generalizing the wrapper so that it can be adapted to various input structures, it is necessary to take care not to reduce the accuracy of extraction because spurious information may be accepted by generalization.

In general, representations of an attribute and its values included by input data differ depending on information sources. Moreover, some attributes include missing values and others include collections of values but not a simple value. In case two or more attributes are involved as in tables, there may be differences in the order of the attributes as well. All these make information extraction more complex.

Moreover, it is necessary to keep the cost of development and that of maintenance as low as possible from a practical viewpoint.

First the wrapper segments input data into two or more attributes. Next it extracts the values of the attribute by applying rules for attribute extraction to each of the attributes and unifies them to entities. If the task of the wrapper also includes any preprocessing, it will consist of the following subtasks (see Fig. 12.6).

1. Crawl input data from one or more data sources.
2. Learn extraction rules by using input data as training data or create extraction rules manually based on inspections of input data.
3. Extract values by applying the extraction rules and output the values in appropriate forms.

In other words, information extraction itself may be an application of data mining techniques.

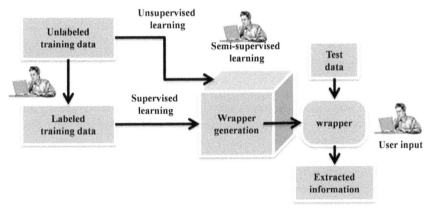

Figure 12.6 Methods for generating information extraction wrappers.

12.2.3 Approaches to Information Extraction

Both creating extraction rules either by machine learning or manually and out putting the extracted values in appropriate forms are the most important among the subtasks of information extraction. The outline of some approaches to these issues will be explained based on the survey work [Chang et al. 2006] as follows.

(1) Manual methods

The methods in this category, where wrappers are manually created, include the following examples.

TSIMMIS [Hammer et al. 1997] enables the user to describe patterns matching with input values and associated processing by commands of a special form [a variable, source, and pattern] and to output extracted data in the OEM form, which Data Guides use so as to describe semi-structured data.

Minerva [Crescenzi et al. 1998] allows the user to describe regular expressions. The system performs pattern matches using them and outputs values by executing procedures as the exception handler associated with the matched patterns.

WebOQL [Arocena et al. 1998] is a search language based on the data model called Hypertree, which can handle HTML, XML, or nested relations. The user issues a query against pages by using WebOQL and extracts values as the result.

(2) Supervised methods

The methods in this category create wrappers based on the training data set prepared beforehand by humans. They include the following examples.

SRV [Freitag 1998] generates the logical rules which judge whether an input fragment is a target of extraction in order to extract the value of a single attribute. SRV tries to learn rules that can detect as many positive examples and as few negative examples as possible by a greedy approach.

RAPIER [Califf et al. 1998] learns patterns which treat a single attribute using both syntax and semantics.

WHISK [Soderland et al. 1999] outputs patterns which extract two or more attributes using regular expressions, based on the training data prepared manually. First WHISK constructs general rules and then specializes them gradually.

STALKER [Muslea et al. 1999] models semi-structured documents as input by a tree structure called Embedded Catalogue. The leaf node of the tree is an attribute to extract and the nonleaf node is a list of tuples. In a hierarchical manner along with the tree structure, STALKER applies rules which extract a child from the parent and those which divide the list into tuples so as to extract data.

(3) Semi-supervised methods

The methods in this category obtain a rough example from the user and generate extraction rules. As the input data may not be strictly correct, the user needs to post process extraction rules generated by the methods.

In OLERA [Chang et al. 2004], the user shows the system a place from which values should be extracted and then the system creates common patterns which can extract other similar values based on the edit distance of the character strings.

In Thresher [Hogue et al. 2005], the user specifies semantic contents and their meaning. The system creates wrappers based on tree edit distances, too. Furthermore, the user can associate a wrapper node with an RDF class and a proposition as its meaning.

(4) Unsupervised methods

Fully Automatic generation of wrappers is performed by the methods in this category.

RoadRunner [Crescenzi et al. 2001] constructs the wrapper of pages by reasoning the grammars of HTML documents, assuming the process of creating the Web site to be that of making the HTML documents from the back-end databases at the sites.

EXALG [Arasu et al. 2003] infers templates instead of grammars.

The generation systems based on the above four kinds of information extraction wrappers are collectively illustrated in Fig. 12.6.

12.3 Deep Web Mining

Web sites which are manually created in advance and are mainly composed of texts are called surface Web or shallow Web. On the other hand, Web sites which have a dedicated database or repository at the back end of the Web server to store a large amount of data and dynamically make pages of search results matching with the user's search terms like Amazon or Google, are called hidden Web or deep Web. In that sense the deep Web are also called Web databases. The data of the deep Web are 500 times as large as the data of the surface Web and continue to increase rapidly [He et al. 2007].

As mentioned above, the latest commercial websites and social websites are mostly the deep Web, which have a database management system at the back end. However, the mining methods explained so far, especially crawling methods, are insufficient for these Web sites.

According to the different purposes for mining the deep Web, i.e., to collect data from the deep Web or to understand the meaning of the deep Web, the issues to solve differ as follows:

(Purpose 1) Collect data from the deep Web.
1. Discover the deep Web services.
2. Extract terms (i.e., conditions) which should be fed into the input form for querying the database.
3. Select the terms that can be used for syphoning the database from pages as results of the queries. The image of syphoning is illustrated in Fig. 12.7.
4. Attain the required ratio of the coverage of the syphoned result over the whole original database.

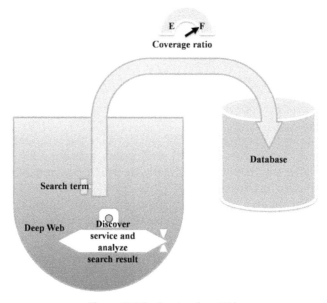

Figure 12.7 Syphoning deep Web.

(Purpose 2) Understand the meaning of the deep Web.
1. Discover the deep Web services.
2. Understand the structure and meaning of the query form.
3. Choose terms (i.e., conditions) which should be fed into the form.
4. Understand the structure and meaning of the pages as a result of the query for querying the database.
5. Analyze and describe the deep Web services as a whole or build information extraction wrappers for the deep Web.

An important issue common to the above two purposes is discovery of deep Web services. Discovery of query forms is mandatory for the issue. It is possible to use the following conditions as heuristics for that purpose.

- A page has GET and POST.
- There are two or more fields. However, neglect fields for free search terms.
- A page has no field that asks for the input of personal information, such as ID, password, and credit card number.

In the structural analysis of forms, annotations to forms (i.e., interpretations) are created through the following processes.

1. Search for tags, name fields, or ID fields.
2. Utilize texts in front of fields.

3. Create candidate consistent annotations to forms by matching them with domain concepts (i.e., ontology).

In the case of purpose one, data with high frequency are collected and used search conditions for the sites. Each condition is specified at the time of search and then the condition that returns a large number of data (namely, the condition with a high ratio of coverage) is chosen. In this case, one of the research issues to solve is collecting as many data as possible under the given constraints, i.e., conditions of holding down the costs required for obtaining the search results within the limits of the resources available for crawlers. For example, the work [Madhavan et al. 2008] is among such approaches.

In the case of purpose two, it is possible to automatically build the wrappers for information extraction by paying attention to any structural repetitions appearing in the result pages. The work [Senellart et al. 2008] is among such approaches.

Conventionally, information extraction called scraping has been used so as to utilize the result of search services. Nowadays, however, Web Service API which can return the result of a fixed number of data in the specified form (e.g., XML) are available. Discovery and use of such API can make mining the deep Web easier.

The following can be considered as applications of deep Web mining.

- Meta search: For the construction of meta search engines using two or more search engines, extract pages (URL) from SERP pages returned by each search engine and unify them.
- Comparison: Collect objects (for example, items and services) belonging to a specific category from two or more deep Web sites of the same kind and compare the values of the same attribute (for example, price) about the same object from all the sites.
- Integration (based on keys): Obtain objects (for example, an author, a paper, and a meeting) belonging to related different categories from the deep Web sites corresponding to the categories. Discover candidate join keys such as an author's name from these objects and join these objects based on the join key and present them to the user.
- Integration (based on non-keys): Obtain objects (for example, data predicting spread of nuclear radiation and precipitation data) belonging to different categories from the deep Websites for the categories. Join and present these objects using the proxy information such as time and space associated with these objects as general join conditions. Here, in particular, the deep websites built from the databases collected by observations or calculations according to scientific disciplines (i.e., explicit knowledge) specific to domains are called collective intelligence Web. It is possible to obtain interdisciplinary collective intelligence

(for example, high level contamination risk area) by using Web sites of heterogeneous collective intelligence in an integrated manner. Such applications of the deep Web in e-science will increase in number and variation from now on.

According to the above described applications, Web databases can be modeled as follows:

- Meta search: Each search engine selects pages as a subset from one virtual database containing all the pages on the Web by crawling and sorts them based on its own metric (PageRank and HITS). However, the rank of each page differs from one search engine to another. Therefore it is necessary for meta search engines to globally rank pages.
- Comparison: Each deepWebsite selects a subset from one virtual database containing all the objects belonging to the same category by a certain method and stores the subset by inserting a new value or updating the existing value of a certain attribute of the objects. For example, the price of the same commercial item may differ from one shopping site to another. Therefore, it is necessary for comparison sites to sort the objects in order of certain attributes such as price and star rating so that the user can find wished for items easily.
- Integration (based on keys): Each Web site contains a database (or a subset) of the objects which belong to its own category. Furthermore, the attributes of each database contain what can generally be used as a key of join with other databases. For example, scientific journal papers and international conference papers written by a certain author are published at the databases of the scientific journals and the international conferences, respectively. They can be joined by the author name as the join key.
- Integration (based on non-keys): Each web site contains a database (or a subset) of the objects which belong to its own category. Furthermore, the attributes of each database contain temporal and spatial information, which can be universally used as a condition of join with other databases. For example, if an overlap is calculated between predicted diffusion of nuclear radiation by SPEEDI (System for Prediction of Environmental Emergency Dose Information) and actual precipitation (quantity of rain clouds) within the same temporal interval and spatial region and the result is displayed on a single map, high level contaminated areas can be roughly predicted. Integrated analysis of images of a rain radar (changed from the image created by Dr. Kitamoto [Kitamoto 2011]) and result images of SPEEDI (changed from the image created by the Nuclear and Industrial Safety Agency

[Nuclear and Industrial Safety Agency 2011]) about the same time and location can predict high-level radioactively contaminated areas as interdisciplinary collective intelligence. In Fig. 12.8, the outline of the structure for integrated analysis is illustrated.

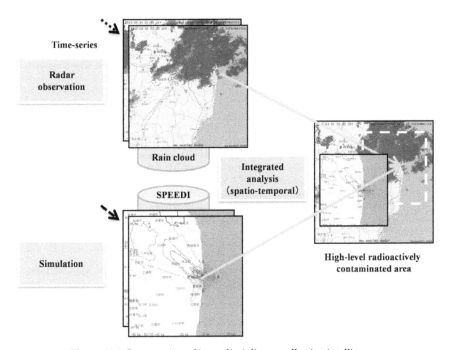

Figure 12.8 Construction of inter-disciplinary collective intelligence.

Color image of this figure appears in the color plate section at the end of the book.

References

[Arasu et al. 2003] A. Arasu and H. Garcia-Molina: Extracting structured data from Web pages. In Proc. of the ACM SIGMOD Intl. Conf. on Management of Data, pp. 337–348 (2003).

[Arocena et al. 1998] G.O. Arocena and A.O. Mendelzon: WebOQL: Restructuring documents, databases, and Webs. In Proc. of the 14th IEEE Intl. Conf. on Data Engineering, pp. 24–33 (1998).

[Califf et al. 1998] M. Califf and R. Mooney: Relational learning of pattern-match rules for information extraction. In Proc. of AAAI Spring Symposium on Applying Machine Learning to Discourse Processing (1998).

[Chang et al. 2004] C.-H. Chang, C.-N. Hsu and S.-C. Lui: Automatic information extraction from semi-Structured Web Pages by pattern discovery. Decision Support Systems Journal 35(1): 129–147 (2003).

[Chang et al. 2006] C.-H. Chang, M. Kayed, R. Girgis and K.F. Shaalan: A Survey of Web Information Extraction Systems. IEEE Trans. on Knowledge and Data Engineering 18(10): 1411–1428 (2006).

[Crescenzi et al. 1998] V. Crescenzi and G. Mecca: Grammars have exceptions. Information Systems 23(8): 539–565 (1998).

[Crescenzi et al. 2001] V. Crescenzi, G. Mecca and P. Merialdo: RoadRunner: towards automatic data extraction from large Web sites. In Proc. of the 26th Intl. Conf. on Very Large Database Systems, pp. 109–118 (2001).

[Freitag 1998] D. Freitag: Information extraction from HTML: Application of a general learning approach. In Proc. of the Fifteenth Conf. on Artificial Intelligence (1998).

[Fu et al. 1999] Yongjian Fu, Kanwalpreet Sandhu and Ming-Yi Shih: A Generalization-Based Approach to Clustering of Web Usage Sessions. In Proc. of WEBKDD, pp. 21–38 (1999).

[Hammer et al. 1997] J. Hammer, J. McHugh and H. Garcia-Molina: Semistructured data: the TSIMMIS experience. In Proc. of the 1st East-European Symposium on Advances in Databases and Information Systems, pp. 1–8 (1997).

[He et al. 2007] Bin He, Mitesh Patel, Zhen Zhang and Kevin Chen-Chuan Chang: Accessing the deep web. CACM 50(5): 94–101 (2007).

[Hogue et al. 2005] A. Hogue and D. Karger: Thresher: Automating the Unwrapping of Semantic Content from the World Wide. In Proc. of the 14th Intl. Conf. on World Wide Web, pp. 86–95 (2005).

[Ishikawa et al. 2003] Hiroshi Ishikawa, Manabu Ohta, Shohei Yokoyama, Takuya Watanabe and Kaoru Katayama: Active Knowledge Mining for Intelligent Web Page Management. Lecture Notes in Computer Science 2773: 975–983 (2003).

[Kitamoto 2011] Asanobu Kitamoto: An archive for 2011 Tōhoku earthquake and tsunami. http://agora.ex.nii.ac.jp/earthquake/201103-eastjapan/Accessed 2011 (in Japanese).

[Liu 2007] Bing Liu: Web Data Mining–Exploring Hyperlinks, Contents, and Usage Data. Springer (2007).

[Madhavan et al. 2008] J. Madhavan, D. Ko, L. Kot, V. Ganapathy, A. Rasmussen and A. Halevy: Google's Deep-Web Crawl. PVLDB 1(2): 1241–1252 (2008).

[Mobasher et al. 2000] B. Mobasher, R. Cooley and J. Srivastava: Automatic Personalization Based on Web Usage Mining. CACM 43(8): 142–151 (2000).

[Muslea et al. 1999] I. Muslea, S. Minton and C. Knoblock: A hierarchical approach to wrapper induction. In Proc. of the Third Intl. Conf. on Autonomous Agents (1999).

[Nuclear and Industrial Safety Agency 2011] Nuclear and Industrial Safety Agency: A result of SPEEDI. http://www.nisa.meti.go.jp/earthquake/speedi/speedi_index.html Accessed 2011 (closed as of this writng)

[Senellart et al. 2008] Pierre Senellart, Avin Mittal, Daniel Muschick, Remi Gilleron and Marc Tommasi: Automatic wrapper induction from hidden-web sources with domain knowledge. In Proc. of the 10th ACM workshop on Web information and data management, pp. 9–16 (2008).

[Shahabi et al. 1997] Cyrus Shahabi, Amir M. Zarkesh, Jafar Adibi and Vishal Shah: Knowledge Discovery from Users Web-Page Navigation. In Proc. of IEEE RIDE, pp. 20–29 (1997).

[Soderland et al. 1999] S. Soderland: Learning to extract text-based information from the world wide web. In Proc. of the third Intl. Conf. on Knowledge Discovery and Data Mining, pp. 251–254 (1997).

[Tan et al. 2002] Pang-Ning Tan and Vipin Kumar: Discovery of Web Robot Sessions based on their Navigational Patterns. Data Mining and Knowledge Discovery 6(1): 9–35 (2002).

[Yan et al. 1996] Tak Woon Yan, Matthew Jacobsen, Hector Garcia-Molina and Umeshwar Dayal: From User Access Patterns to Dynamic Hypertext Linking. WWW5/Computer Networks 28(7-11): 1007–1014 (1996).

Media Mining

Social big data are characterized not only by the volume of the data but also by the variety of the data structures. Advanced mining targeted at XML data, trees, and graphs, multimedia mining targeted at images and videos, and stream mining targeted at time-series data will be explained in this chapter.

13.1 Mining XML

13.1.1 XML Mining

This subsection explains mining semi-structured data such as XML. Although there are an increasing number of individual researches about mining XML in recent years, there have been very few systematic reviews of XML mining. Following the classification of Web mining, XML mining would be classified as follows:

- XML structure mining
- XML contents mining
- XML access log mining

Next, supposing it is the main objective of XML mining to discover frequent patterns in XML data, the following applications seem promising. In this subsection, no strict distinction between XML data and XML documents is made.

- Effective storage of XML data using relational databases
  By summarizing frequent structures of XML data in relational databases, the number of join operations of tables (i.e., relations) can be reduced and queries can be efficiently processed.

- Assistance in formulating XML data queries and views
  In general, a query cannot be formulated unless structures of XML data are known in advance. If structures of frequent data are understood,

queries corresponding to such data can be described using the structures. Furthermore, if such queries are defined as XML data views, they can be reused.

- Indexing to XML data
  XML data queries can be efficiently processed if indices are built for frequent access to XML data in advance.

- Summarization of XML data
  Subdocuments which frequently occur in an XML document may summarize the whole document. In case of the structures only, they represent the outline of the whole document.

- Compression of XML data
  Efficient compression of XML data can be performed using structures and contents which occur frequently.

- Extraction of access patterns to Web pages
  Access patterns which occur frequently are extracted from Web access logs and used for recommendation of the Web pages or redesign of the Web sites. In general, as the access patterns can be more naturally modeled by tree structures or graph structures than by linear lists, XML data can be used for representation of the access patterns.

Below, XML structure mining, XML contents mining, and XML-enabled access log mining will be explained.

13.1.2 XML Structure Mining

XML structure mining mainly identifies the structures of XML documents from the viewpoints of hierarchies and attributes of elements. It can be further classified into mining structures within an XML document and mining structures between XML documents. The former and the latter are called intra-XML structure mining and inter-XML structure mining, respectively.

After explaining association analysis, cluster analysis, and classification of XML data, which are applications of the fundamental techniques of data mining, individual techniques will be explained in detail.

(1) Intra-XML structure mining

It is possible to discover the relationships between tags within an XML document by applying association rule mining to the XML document. For example, the hierarchical structures of XML document are transformed into linear transaction data and combinations of tags which frequently appear together are discovered by paying attention to the inclusive relationships of elements or tags within the same elements, respectively. It is possible to distinguish homonyms of tags, for example, by the differences in other tags which appear simultaneously. In addition to classification by the Naive Bayes method, there are classification techniques based on dictionaries and thesauri. Clustering based on the EM method can be used for generalization of meanings of tags [Manning et al. 1999].

(2) Inter-XML structure mining

Inter XML document structure mining is related to discovery of relationships between objects on the Web such as themes, organizations, sites and to discovery of relationships between elements of an XML document set. Generally, while intra-XML structure mining is targeted to a single name space, inter-XML structure mining is concerned with two or more name spaces and URIs.

Like intra-XML structure mining, inter-XML structure mining can be classified as follows:

- In order to discover relationships between tags in two or more XML documents, it is possible to apply association analysis.
- In classification of a set of XML documents, given DTD can be viewed as classification rules. If a new XML document is given, the document is assigned to a document class corresponding to its relevant DTD. In other words, a class to which the given document will be assigned is determined by validating it against DTD of the document class.
- Clustering a set of XML documents requires to discover the similarity of various XML documents. If two or more DTD are given, XML documents can be grouped based on the similarity between such DTD. For the group made in this way, a new DTD which is super to DTD of the documents belonging to each group will be made.
- In general, the producers and the users of DTD (i.e., maker of individual XML documents) are considered to be separate. Such relationships can be viewed as those in Web structure mining such as the authority and hub in HITS.
- It is possible to predict schema structures (i.e., models) of XML from XML documents (i.e., instances), based on observation of structures of XML instances such as elements or attributes. This can be used as preparation for performing efficient storage and search of XML instances in XML databases. Furthermore, it enables to predict structures of XML documents as results generated from a query of XML documents.

As mentioned above, if structure mining spanning two or more XML documents is broadly perceived, it will be a technology which overlaps with Web structure mining or Web contents mining. However, inter-XML structure mining differs from Web structure mining in that the former is focused more on the internal structures of XML documents. However, if XML-conformed HTML (namely, XHTML) comes to be more widely used, it is thought that differences between the two technologies will be reduced increasingly.

As enabling technologies of such inter-XML structure mining, some technologies depend on the existence of DTD or schemas and others do not. The former includes the technique [Shanmugasundaram et al. 1999] and the latter includes DataGuides [Goldman et al. 1997]. Here, the latter technology is considered to be more useful because it is not generally guaranteed that DTD are available nor that documents conform to DTD even if any.

Hereinafter, outline extraction of XML, automatic generation of DTD, and discovery of schemas for efficient storage will be explained as individual technologies as to XML structure mining. Classification and clustering based on the structures of XML will be briefly touched on at the end of this subsection.

(a) Outline extraction

Here, DataGuides will be explained as one of the technologies applicable to outline extraction of XML data structures. DataGuides create the summary (i.e., outline) about the structures of semi-structured data such as XML data. It aims at enabling the user to browse the structures of XML databases or formulate queries against XML databases. Moreover, it helps the system to create indices for efficient access to XML databases.

OEM databases will be introduced as semi-structured databases which DataGuides are targeted at. Objects, as components of OEM databases, are uniquely identified by object identifiers. An object is either a primitive value (e.g., character string, numerical value, image, and program) or it is composed of zero or more subobjects. Such objects and subobjects are connected by links together with labels. This can be considered a model of XML databases. An example of an OEM database is illustrated in Fig. 13.1.

Before defining a DataGuide, a label path is defined as follows.

(Definition) A label path, a data path, an instance, and a target set
- A label path is a sequence of labels along a path.
- A data path is a sequence of a pair of a label and an object along a path.
- If a label sequence of a data path *d* is equal to that of a label path *l*, then the data path *d* is an instance of the label path *l*.
- A target set is a set of objects which can be reached by traversing a label path.

A DataGuide is defined using the above concept as follows.

(Definition) DataGuide
A DataGuide object *d* for an OEM source object *s* satisfies the following conditions.

- Every label path of *s* has exactly one instance in the data path of *d*.
- Every label path of *d* is a label path of *s*.

For example, a DataGuide corresponding to the OEM database of Fig. 13.1 is illustrated in Fig. 13.2.

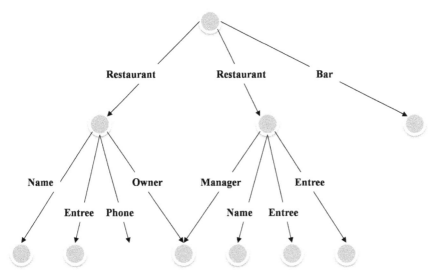

Figure 13.1 OEM database.

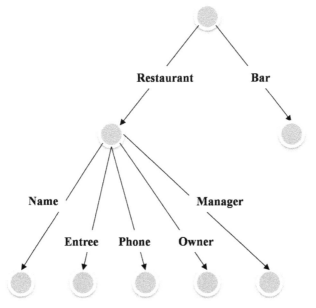

Figure 13.2 DataGuides.

Furthermore, the concept of a strong DataGuide is defined as follows.

(Definition) Strong DataGuide
In a strong DataGuide, each DataGuide object corresponds to the target set of all label paths that reach that DataGuide object.

For example, while Fig. 13.3b is a strong DataGuide object for Fig. 13.3a, Fig. 13.3c is not.

Similar to *Query By Example* (QBE) [Zloof 1977], the user can formulate and execute a query based on graphically presented DataGuide objects. DataGuide objects also provide indices to paths, which can be used for query optimization.

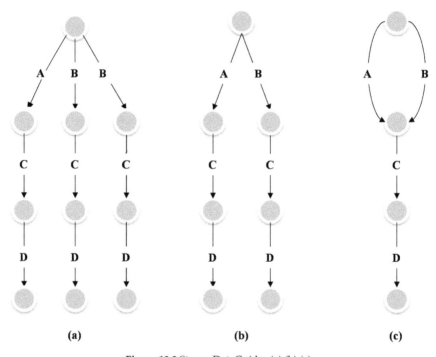

(a)   (b)   (c)

Figure 13.3 Strong DataGuides (a) (b) (c).

(b) Automatic generation of DTD

Although DataGuides provide information on the structures of semi-structured data such as XML, creation of DTD is not its original purpose. Some researches which aim to automatically generate DTD from XML data will be introduced here. They include XTRACT [Garofalakis et al. 2000] and DTD-miner [Moh et al. 2000]. If a collection of XML data is given, these methods will generate DTD based on the collection.

Here XTRACT will be explained. XTRACT consists of the following steps.

(Algorithm) XTRACT
1. Generalization: Using empirical rules, find a partial sequence with high frequency from an input sequence and let it be a DTD candidate.
2. Factoring: Compute an intersection of DTD candidates as results obtained at the previous step and make it a DTD candidate anew.
3. Selection based on the MDL principle [Tan et al. 2002]: Choose DTD with the minimum MDL value from DTD candidates obtained at the previous step.

For example, let us compute the MDL cost for DTD $(a|b)^*$ to input sequences {ab, abab, ababab}. Description of DTD requires 6 characters. Each of the three sequences require 2, 4, and 6 characters as to specify choice of either a or b, respectively. The number of repetitions in each sequence equally requires one character. On the whole, $6 + 1 + 2 + 1 + 4 + 1 + 6 = 21$ characters are needed. Therefore, the MDL cost in this example is 21.

Please note that it is not necessarily possible to use DTD as schemas of XML databases as they are even if there exist DTD as specification for structures of XML documents. Thus, the original purpose of DTD is to describe the structures of XML instances. Therefore DTD lack information necessary for the design of database schemas appropriate for efficient storing or efficient query processing.

Some researchers have aimed to build a normalization theory in XML databases [Wu et al. 2001] or XML functional dependency [Chen et al. 2003] of XML schemas, based on DTD as solutions to these issues. Others have proposed methods to determine database schemas directly based on instances without passing through DTD. One of the latter methods will be described in the next subsection in more detail.

(c) Discovery of storage structures

The techniques of storing XML data using relational databases which have been proposed so far will be classified into the following two.

- The technique of storing XML data in databases with the schemas pre-determined by the system.
- The technique of storing XML data in databases with the schemas dynamically generated either from models such as DTD or instances.

First, the techniques of storing XML data in databases with the schemas prepared in advance will be described. Here let us assume that elements and inclusion relationships between them correspond to nodes and edges of trees, respectively. Florescu and others [Florescu et al. 1999] have proposed a method to store edges and values in separate tables, a method to use a separate table for each kind of edges, and a method to store all elements in universal relations, assuming that all elements have values and edges. Jiang and others [Jiang et al. 2002] and Yoshikawa and others [Yoshikawa et al. 2001] have proposed methods to store XML data by dividing them into values (i.e., character data), elements, and paths.

These techniques have the following advantages.

- Any XML data can be stored in databases of the same schemas.
- XML data with various structures can be stored in consideration of order of elements.
- Queries including parent-child relationships can be easily converted into SQL.

However, these have the following disadvantages with respect to flexibility.

- Data types (for example, a numerical value and a character string) according to each element cannot be freely defined.
- Data aggregation such as sum and average in XML data becomes rather complicated as compared with the techniques of storage by dynamic schemas.

On the other hand, the techniques of storage based on dynamic schemas enable the system to flexibly define data types, such as a numerical type and a character string, according to a value of each element. However, since the number of tables will also increase as the kind of elements increases, join operations of tables requiring large computational cost will occur frequently in queries. As a result, there occurs a risk of increase in the response time.

One of solutions to this issue is reduction of the number of tables involved in join operations. Some researchers such as [Deutsch et al. 1999] and [Shanmugasundaram et al. 1999] proposed methods to generate schemas by using DTD as researches along such a direction. Although DTD are indispensable to these techniques, end users cannot necessarily describe DTD. The technique of assisting schema generation of an object relational database (ORDB) is proposed based on DTD, the number of occurrences of XML elements, and frequent queries [Klettke et al. 2001]. This, however, does not take separating tables for efficiency into consideration.

Furthermore, DTD lack information sufficient to define data types and to determine the maximum number of occurrences of child elements of an element since the primary purpose of DTD is not to describe relational database schema. Therefore DTD is not sufficient to generate efficient database schemas. That is, XML data obtained even from the same DTD tend to become various and generally it is not necessarily possible to uniquely determine efficient database schemas for such XML data.

On the other hand, XML Schema, which has a power of expression higher than that of DTD [XML Schema 2014], contains much information applicable to database schema generation. So a technique of generating database schemas using such information seems promising. However, it is also difficult for end users to directly define XML Schema like DTD.

Then, the author and others [Ishikawa et al. 2007] have devised a technique of generating database schemas only by using statistical analysis of XML documents so as to process XML documents without models such as XML Schema and have shown that the technique is more efficient for processing queries compared with the method of dividing tables by using normalization only (i.e., baseline method).

(d) Classification and clustering based on structures

As classification based on the structures of XML data, XRules uses tree structure mining [Zaki et al. 2003]. Moreover, some works such as [Chawathe 1999] and [Dalamagas et al. 2004] define and use similarity of XML data based on the edit distances between tree structures so as to cluster XML data. XProj [Aggarwal et al. 2007] performs clustering based on association rule mining for series data. A method [Harazaki et al. 2011] defines and uses similarity of ordered sets which have XML paths as elements in order to cluster XML data.

13.1.3 XML Content Mining

In this subsection, the outline of the content mining of XML data will be explained. XML content mining is a task similar to Web content mining and general text mining since it is principally aimed at values (i.e., character strings) surrounded with tags. However, XML content mining also pays attention to the structures of tags surrounding the values as well. Therefore it differs from text mining, which focuses on values themselves.

XML content mining is further classified into content analysis and structure clarification. In addition, since compression of XML data also observes both structures and contents of XML data, it will also be explained here.

(1) Content analysis

In classification of XML documents, if DTD of documents are known beforehand, it is possible to narrow search space and to reduce the processing time as to comparison by considering only documents which conform to DTD corresponding to a certain class (i.e., category). If DTD are similar to each other, sets of values of corresponding elements are also similar with high possibility. Therefore, in clustering XML documents, search space can also be reduced using the similarity of DTD. Conversely, it is possible to discover synonymous tags of DTD using the similarity of the contents. On the other hand, polysemy as to values may sometimes cause problems. In such a case it is expected that tags surrounding the values may contribute to removal of ambiguities.

(2) Structure clarification

If XML instances with different DTD are assigned to the same cluster, it will lead to discovery of semantic relevance between these DTD. Conversely, if XML instances with the same DTD are assigned to different clusters, it may be necessary to suspect the presence of polysemy as to tags of DTD.

As a tool of XML content mining, the query languages for XML data such as XQuery [XQuery 2014] become important.

(3) Compression of XML data

As XML data are self-descriptive, they are essentially redundant. Although redundancy can be an advantage in some aspects, it may have a bad influence on performance as to the exchange and storage of XML data through networks. Therefore, a lot of techniques of compressing XML data efficiently have been studied using the structures and the contents of XML data. Generally, processing speed, compression ratio, and reversibility are considered to be important in compression algorithms.

Furthermore, it must be taken into consideration whether queries about compressed XML data are allowed or not. Methods of XML data compression are divided roughly into two groups according to whether such queries are possible or not. The former group includes XGrind [Tolani et al. 2002], XPRESS [Min et al. 2003], the system of the author and others [Ishikawa et al. 2001]. The latter includes XMill [Liefke et al. 2000].

Here, from the viewpoint that systems allowing querying compressed data are more useful, the system of Yokoyama and others will be described. The system uses a method similar to Huffman encoding. That is, the frequencies of tags are counted and tags are sorted in ascending order of frequency. First the shortest code is assigned to the tag with the highest frequency, Next the second shortest code is assigned to the tag with the second highest frequency. This process continues until all tags are assigned codes. However, XML contents (i.e., values) themselves are not encoded. XML data compressed by this way still remain to be XML data as the original data are. Therefore, transparent access can be made to compressed XML data only by attaching wrapper programs for the existing XML tools that translate between the original tags and encoded tags and check of conditions on values can be done without decompressing compressed data. XGrind also uses Huffman coding for compressing XML data, preserving the structures of the original XML data.

(2) Automatic generation of XML data

Once a lot of XML data have been accumulated, the necessity of performing the quality assessment as to the scalability of services or systems for such data will emerge. It is desirable that real data can be used for such assessment. However, it is actually difficult to obtain real data of various sizes required by performance analysis. To synthesize XML data or to make artificial data seems promising as a solution to this problem.

   Approaches to generation of artificial data are roughly divided into two as follows.

- Generate XML data so as to conform to user-specified data structures. For example, xmlgen [Aboulnaga et al. 2001], ToXgene [Barbosa et al. 2002], and Cohen's method [Cohen 2008] are included in this category. However, structures which can be specified are fixed or restrictive in some systems. Although structures can be specified by DTD in other systems, values have restrictions.
- Generate XML data so as to reflect the statistical features of real data. For example, XBeGene [Harazaki et al. 2011], analyzes real data given as input, extracts the structures and statistical features (e.g., frequencies of elements and values), and generates data of arbitrary size which reflect them. XBeGene expresses the extracted structure using DataGuides. Further, XBeGene can generate data which satisfy queries specified by the user.

13.2 Mining More General Structures

Now mining technology of tree structures and graph structures, which are more general than XML, will be explained below.

(1) Tree structure mining

Here, according to Zaki's work [Zaki 2002], after the problem of mining frequent trees and the related concepts will be defined, concrete algorithms will be explained.

   (Definition) Mining frequent trees
- When the tree data set D and the minimum support $minSup$ are given, mining frequent trees is finding s included in the tree (transaction) t as each element of D whose support (frequency in D) is equal to or more than $minSup$. A set of trees s which consist of k branches is denoted by F_k.

   (Definition) Scope
- Let the right most node within a partial tree whose root is the node n_l be n_r. Then the scope of the node n_l is denoted by $[l, r]$.

- A scope list is a list whose element is a pair of a node within the tree and its scope.

(Definition) Prefix
- A prefix of a tree expresses a sequence of labels of nodes of the tree traversed in preorder. Here, backtracking (i.e., return) to the parent of a node is denoted by -1. For example, the tree T and its prefix are illustrated in Fig. 13.4a. A scope is also added to each node.

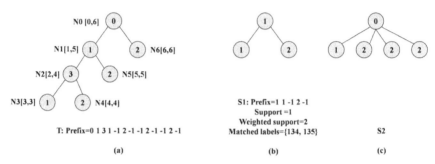

Figure 13.4 Prefix and partial tree.

(Definition) Embedded partial tree
- That the tree s is contained in the tree t means that all the nodes of s are also those of t and n_x is an ancestor of n_y in t if n_x is a parent of n_y in s, that is, for all the branches (n_x, n_y) of s. In that case, it is said that the tree s is an embedded partial tree of the tree t. From now on, unless otherwise noted, it is called a partial tree for short.

For example, the partial tree S_1 of the tree T is illustrated in Fig. 13.4b, together with the supports, weighted supports, and matched labels.

(Definition) Equivalence class $[P]_k$
- If two k-partial trees X and Y have a common prefix P up to $(k–1)$-th node, X and Y become members of an equivalence class. The equivalence class in this case is denoted by $[P]_k$.
- A prefix (i.e., partial tree) which is created by adding to P an element (x, i), i.e., the label x given to the position i, is called $P_{x'}$.

An example of join (×) operation of an equivalence class and a partial tree is illustrated in Fig. 13.5. Furthermore, an example of join ($\cap_x$) operation of scope lists is illustrated in Fig. 13.6.

Now, an algorithm for discovering frequent trees using these concepts will be introduced.

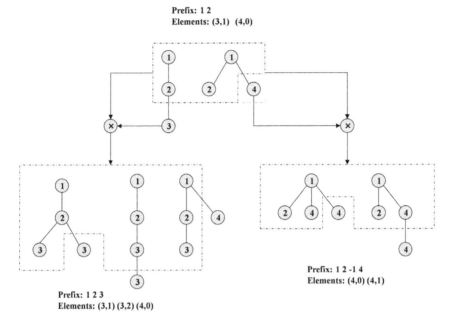

Prefix: 1 2
Elements: (3,1) (4,0)

Prefix: 1 2 3
Elements: (3,1) (3,2) (4,0)

Prefix: 1 2 -1 4
Elements: (4,0) (4,1)

Figure 13.5 Join of equivalence class and partial tree.

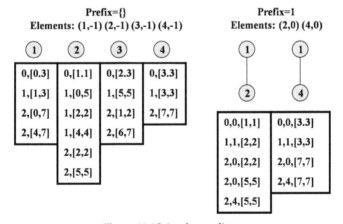

Prefix={}
Elements: (1,-1) (2,-1) (3,-1) (4,-1)

Prefix=1
Elements: (2,0) (4,0)

Figure 13.6 Join of scope lists.

(Algorithm) TreeMine (*D, minSup*)
1. $F^1 \leftarrow$ all frequent 1-partial trees;
2. $F^2 \leftarrow$ all frequent 2-partial trees belonging to $[P]_1$
3. For all $[P]_1$ do the following {
4.   Enumerate_Frequent_Subtree ($[P]_1$);};

(Algorithm) Enumerate_Frequent_Subtree ([P])
1. For each element (x, i) belonging to $[P]$ do the following {
2. $[P_x] \leftarrow \varnothing$;
3.   For each element (x, j) belonging to $[P]$ do the following {
4. $R \leftarrow \{(x, i) \times (y, j)\}$;
5. $L(R) \leftarrow \{L(x) \cap_x L(y)\}$;
6.   if (r in R is frequent) then
7. $[P_x] \leftarrow [P_x] \cup \{r\}$;};
8.   Enumerate_Frequent_Subtree($[P_x]$);};

Zaki applied a tree structure mining technology to Web access log mining as an application of XML mining. This is called XML access log mining. Zaki mined the access logs by using the following three data structures for comparison.

(i) Access page set
(ii) Access page sequence
(iii) Access page tree

They observed that although processing time increased in order of (i) < (ii) < (iii) as a result, the amount of information acquired also increased in the same order.

Furthermore, Zaki and others extended TreeMiner so as to discover frequent XML trees by specifying a separate minimum support for every class and created a structural classifier named XRules [Zaki et al. 2003] of XML data.

(2) Graph mining

So far mining semi-structured data represented by XML, i.e., ordered trees, has been described. Here, more generally, mining graphs, which have caught the increasing attention of researchers with the spread of social media, will be described.

First, the problem to be solved (i.e., mining frequent graphs) will be defined.

(Definition) Mining of frequent graphs
- When a set D of graphs and the minimum support *minSup* are given, discovery of a graph g isomorphic to partial graphs of a graph (i.e., transaction) t as each element of D which has a support (i.e., frequency in D) more than or equal to *minSup* is called frequent graph mining.

Then the concept of graph isomorphism is defined as follows.

(Definition) Isomorphic graphs
- If there exists a one-to-one mapping between two sets of nodes constituting two separate graphs and the mapping preserves the adjacencies of two nodes in each graph, it is said that the two graphs are isomorphic to one another.

The problem, especially called the problem of isomorphic partial graphs, that decides whether a partial (or sub) graph of one graph and another graph are isomorphic is known to be among so-called NP complete problems, which cannot be solved in polynomial time.

The graph mining algorithm using the Apriori principle like association rule mining will be described. They include FSG [Kuramochi et al. 2001] and IAGM [Inokuchi et al. 2000]. Here, FSG will be introduced.

Fundamentally, the algorithm has the same structure as the Apriori algorithm as follows.

(Algorithm) FSG (D, σ)
1. $F^1 \leftarrow$ all frequent 1-subgraphs of D;
2. $F^2 \leftarrow$ all frequent 2-subgraphs of D;
3. $k \leftarrow 3$;
4. while $F^{k-1} \neq \emptyset$ {
5. $C^k \leftarrow$ fsg-gen(F^{k-1});
6.   For each element g^k *of* C^k do the following{
7. g^k.count$\leftarrow 0$;
8.   For each transaction t of D do the following{
9.   if g^k is included by t then g^k.*count* $\leftarrow g^k$.*count*+1;};};
10. $F^k \leftarrow \{g^k | g^k$.count $>= minSup\}$;
11. $k \leftarrow k+1$;};
12. return $F^1, F^2, \ldots, F^{k-2}$;

fsg-gen (F_k) generates candidate partial graph C_{k+1} from the frequent partial graph F_k. Thus, frequent k-partial graphs which have a $(k-1)$-partial graph in common are combined to generate C_{k+1}. The common partial graph in this case is called core. Only one k-candidate item is made from two $(k-1)$-frequent items in case of sets. In the case of graphs, however, two or more candidate graphs are made as follows.

(i) Nodes with the same label are separate (see Fig. 13.7a).
(ii) The core itself has an automorphic graph (see Fig. 13.7b).
(iii) It has two or more cores (see Fig. 13.7c).

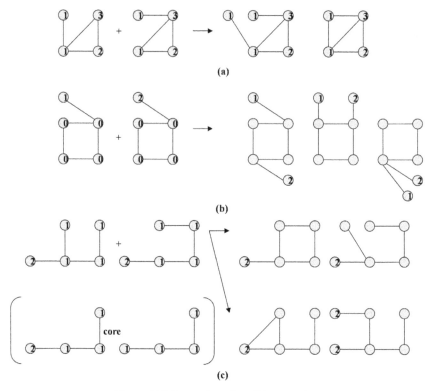

Figure 13.7 Generation of candidate graphs.

A TID list is used for enumeration of candidate graphs like the Partition algorithm already explained as an extension to the Apriori algorithm. Kuramochi and others have used chemical compounds in the experiments, have evaluated the above-described algorithm, and have ascertained that frequent graphs can be discovered at comparatively high speed when the minimum support is small (10% or less).

13.3 Multimedia Mining

Traditionally, most data treated in information systems have been structured data. According to the development of the Web, however, not only texts but also images, videos, and audios are becoming more and more popular. In addition to Flickr and YouTube, Twitter and Facebook also enable the users to include photos and videos in their articles. Therefore it is desirable that so called multimedia data such as texts, images, and videos can be searched as structured data from databases. Text search has been studied over many years in the field of information retrieval, and has reached the practical level in Web search. Similarly, services for searching images and videos through user-specified search terms are popular on the Web. However, it is very difficult for the user to describe all of the contents and the features of multimedia data only by search terms. Therefore, the dedicated mining technology of multimedia is necessary as the foundations of search, classification, and clustering.

Research and development related to images among multimedia have been comparatively advanced. Taking image search as an example, some approaches to media search will be explained below. As a facility for searching images, it is conceivable to use search terms extracted from an anchor text (i.e., text on the link to an image). This approach is currently used in most Web searches. However, such terms alone cannot fully describe all the features of the images. Content-based image retrieval has long been studied based on the features of images, as will be described below.

The main features of images used by the content-based retrieval include the following:

- The histogram of a color: This feature is the simplest and easiest to treat. However, it is unsuitable for making a distinction between different textures or compositions.
- Wavelet: This feature can represent the colors, textures, and compositions of images simultaneously. However, it is not suitable for distinguishing between objects with various sizes and positions contained by separate images.
- Combination of two or more features (e.g., the histogram of colors, the layout of colors, the histogram of edges, and CEDD)[1] [van Leuken et al. 2009]: This approach is widely used. Distance measures for each feature and methods of combining them must be carefully selected.

Furthermore, in addition to content-based retrieval of images, applications of the fundamental mining technology can be considered as follows:

- Association analysis of images
  Association analysis is applicable to relationships between features of images and the other information and those between two or more objects contained by images. In the latter case, it may be necessary to take into further consideration spatial relations (e.g., upper and lower, right and left, an inclusive relation) between objects.

- Classification of images
  Classifiers of objects with respect to the features are learned beforehand. If a new image is given, they can help to classify it to the existing class.

---

[1] Color and Edge Direction Descriptor: The image feature that incorporates color and texture information in a histogram.

There have been broad applications in image recognition and scientific researches.

- Clustering images

  Nowadays, a large amount of images are available on the Web. In displaying the search results, it is possible not only to present them in a traditional list or tiles but also to cluster and present similar images. In the latter case, it is necessary to combine various information such as the positions of images, the surrounding texts, and the links in the original page as well as the features of the images so as to raise the accuracy of clustering.

However, the accuracy of search results also has limitations and there exist constraints on application domains as long as techniques based on only such image features are used. In other words, as it is essentially necessary for the user to express the intention of image search, the system needs to consider not only raw features of images but also the meanings contained by the images so as to correctly interpret the intention.

An increasing number of services that allow the users to annotate images by adding tags to them have been provided on the Web. Further, metadata (i.e., Exif tags) and position information (i.e., geographical tags) describing shooting situations have been attached to an increasing number of images on the Web. As previously described, indexing based on search terms collected from anchor texts alone is limited. On the other hand, tags which the users have arbitrarily added to social data including multimedia data are called social tags in general. Such social tags can be considered to represent some meanings of the social data. Social tags and metadata can also be used in image retrieval, image clustering, and image classification.

Data mining targeted at a set of photos on the map by using metadata such as geotags (i.e., location information), exif data (e.g., camera direction, focal length), and social tags (e.g., landmark name) will be described below.

Shirai and others [Shirai et al. 2013] first calculate the density of photos per unit grid contained by a specified geographical space by using the location information of the photos. Next using the DBSCAN method, they cluster grids next to each other whose density is above the specified threshold and calculate the geographical centroid for each cluster. Then they classify a photo contained by each cluster into an inward photo whose angle of view, calculated by the focal length and camera direction, contains the centroid and an outward photo whose angle of view doesn't. By expanding an area from the centroid of each cluster, if the number of the inward photos exceeds that of the outward photos, they estimate the area to contain a place of interest inside. Lastly they have successfully detected the outline of the landmark (i.e., place of interest) based on grids contained by the estimated area.

The author and others first retrieve a set of photos containing a specified search term (e.g., beach) from Flickr. Using the location information of such photos, they detect grids on the map whose density of photos is high. They have invented an algorithm which connects such grids next to each other and draws lines along the connected grids. By using the algorithm, they have successfully illustrated the real coastlines.

In addition to images, it is also conceivable to search, cluster, and classify real-time stream media data such as videos and audios. However, video streams are not simply a sequence of frames (i.e., still images). Rather they consist of a coherent sequence of frames as units called shot. Content-based video retrieval will be explained by taking a research by the author and others [Ishikawa et al. 1999]. Prior to retrieval, the system divides video streams (i.e., MPEG-2 video) into a sequence of shots, extracts representative frames from the shots, and makes thumbnails by using such representative frames for the shots. In retrieval, the system allows the user to filter a set of videos by specifying semantic tags and allows the user to further select only desired shots by clicking on the thumbnails of videos tiled as a filtered result.

The system uses changes in histograms of luminance so as to detect cuts between shots. An area containing moving objects within a frame as well as its moving direction are obtained by detecting an area where motion vectors of macro blocks have uniform directions. At this time, necessary adjustment of cameraworks (e.g., pan, tilt, and zoom) is performed. Furthermore, the area for the moving object is segmented into smaller color regions and the representative color, centroid, and area sizes are calculated for each region. These kinds of features are stored in dedicated databases associated with moving objects. The system allows the user to specify the shape and color of a moving object as a sample by combining colored rectangles together with the motion direction through the graphical user interface (see Fig. 13.8). Then the system retrieves shots containing the user-specified sample by using quad-trees as a hierarchical multi-dimensional index. Thus the system categorizes motion directions into eight (e.g., up, down, right, left, upper right, and down left) and builds eight quad-trees for the corresponding directions. Then the system calculates the color distances between rectangles of the sample and those of each element in the set of moving objects obtained through the indices and calculates the total sum for each moving object and presents a set of shots containing the corresponding moving objects in ascending order of the values.

In addition, it is possible to use MPEG-7 as a general framework which enables the users to describe various features necessary for video search as metadata.

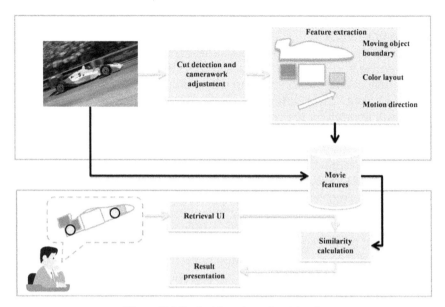

Figure 13.8 Content-based movie retrieval.

Color image of this figure appears in the color plate section at the end of the book.

13.4 Stream Data Mining

Streams are a kind of time series data generated from various sources such as telephone networks, computer networks, sensor networks, and information systems in various fields such as finance, distribution, and manufacturing. Most stream data are tremendously large, changeful, and continuous. As described earlier, social data, especially twitter articles can be viewed as a kind of stream data.

13.4.1 Basic Techniques

Since most stream data are continuously generated and fed into information systems in a rather short interval between successive arrivals, there is a risk of exceeding the throughput of the systems if stream mining is performed in the usual environments which have resource restrictions.

In other words, if the original stream data can be reduced appropriately by selecting some parts from the data or transforming the data into another form, existing technology will be applicable. The following techniques can serve such purposes [Gaber et al. 2005].

- Sampling
  In general, the sampling techniques express the original data by selecting parts from the original data at random and preserving the

characteristics of the original data. In sampling stream data, it is necessary to take any measures against issues such that the size of the whole stream data is unknown beforehand and that the arrival rate of stream data is inconstant.

The sampling techniques can be applied to various applications, which include the estimate of the frequency of the arrival, classification (e.g., the decision tree), clustering (e.g., k-means), and query processing with respect to stream data. The error accompanying sampling is expressed as a function of the sampling rate. Please note that sampling is not suitable for discovery of outlying values.

- Sketching
  Sketching is the technique of sampling stream data on a value-basis and creating the summary of the whole data. Such a sampling technique is especially called vertical sampling. On the other hand, to sample stream data on a time-basis is called horizontal sampling. Sketching can be used for processing aggregation queries and queries against two or more streams. In this case, it is important to guarantee the correctness of a processing result.

- Histogram
  A histogram, a frequency moment, and wavelet can be used as data structures representing the synopsis of stream data. A histogram expresses the frequency of each value. The k-th frequency moment is the sum of the k-th power of the frequency of each value. Thus, if k is equal to zero, one, two, or infinity, it expresses the number of distinctive values, the total frequencies, the variance, or the maximum, respectively. A wavelet develops the original stream data to the total sum of independent basis functions and expresses the original data with the coefficients of such functions.

- Aggregation
  Aggregation literally expresses the quantitative characteristics of stream data by the average and variance, which are statistical concepts.

- Sliding window
  The sliding window is used so as to give more weight to latest data than the prior data. The size of the window expresses the length of the history upon which can be looked back.

- AOG (Algorithm Output Granularity)
  AOG is used to control the output rate of a result in the computing environments which have restrictions in memory and throughput. For example, if available memory is becoming short, the knowledge acquired so far will be merged.

The above are basic techniques used in common with stream mining.

13.4.2 Data Mining Tasks

Below, the technology of stream mining will be seen from a viewpoint of basic data mining tasks (i.e., clustering, classification, and association analysis).

- Clustering
  Clustering streams is required to process data by one path in the limited memory and time.

  STREAM [Babcock et al. 2002] is a clustering algorithm based on k-medians which assigns the N points to the k centers nearest to them. STREAM makes clusters such that the sum of all the squares of the distance between the point belonging to the cluster and its center be minimized. N points are divided into buckets each of which consists of m points and clustering is performed for every bucket. The bucket size is determined such that the bucket fits in available memory. Then it keeps only the set of cluster centers weighted by the number of the members and throws away the other points. If the number of thereby obtained centers exceeds a threshold, clustering will be further applied to them so as to find a set of new centers.

  CluStream [Aggarwal et al. 2003] combines on-line clustering (i.e., micro clustering) and off-line clustering (i.e., macro clustering). Micro clusters are represented by CF in BIRCH extended by the time stamps. On-line clustering first maintains q micro clusters so that they fit in main memory. Each cluster has its own boundary. If data which have newly arrived enter within the boundary of one of the clusters, they are assigned to such a cluster, otherwise a new cluster is made and the data are added to the cluster. In order to fit all the clusters in main memory, macro clustering either deletes a cluster which has passed longest since being created or merges two existing clusters into one according to a certain criterion.

  Macro clustering allows the user to analyze the evolution of clusters. Here let a time horizon be a sequence of stream data. The user can specify h and k as the length of the time horizon and the number of macro clusters, respectively. The system calculates a time horizon by subtracting CF at the time $(t-h)$ from CF at the time t and re-clusters micro clusters contained by the time horizon as individual data so as to get k macro (i.e., higher level) clusters.

- Classification
  Classification of streams must take into consideration both that available memory is usually not enough to rescan all the data and that a model changes with time (i.e., a concept drift occurs).

  First, the Hoeffding bound will be explained. The Hoeffding bound asserts that given the accuracy-related parameter δ, the error between the index r' based on sample data (N items) and the true index r in all the data doesn't exceed the parameter ε defined by the following formula with a probability of ($1-\delta$).

$$\bullet \quad \varepsilon = \sqrt{\frac{R^2 \ln \frac{1}{\delta}}{2N}}$$

Here R is a constant depending on the domain of r. For example, if r is probability, then R is one. If r is the information gain, R is $\log C$, letting C be the number of classes.

Therefore, if the difference of the best index and the second best index is larger than the ε, attributes will be selected based on the former index.

If new data arrives, VFDT (Very Fast Decision Tree) [Domingos et al. 2000] classifies the data by using the current decision tree and stores the data in a leaf. If any leaf has fully accumulated data, the leaf will be expanded as a tree based on the Hoeffding bound. In order to respond to the concept drift issue, CVFDT (Concept-adaptation Very Fast Decision Tree) [Hulten et al. 2001] modifies VFDT so that a decision tree is incrementally made.

Furthermore, it is conceivable that a classifier is created for every chunk of stream data and ensemble learning is performed against the top k classifiers.

- Itemsets counting
Since the basic techniques for counting frequent itemsets are required to scan all the data two or more times, they are inapplicable to itemset counting in stream data. Lossy Counting [Manku et al. 2002] allows the user to specify the minimum support σ and error bound ε. The frequencies f (round numbers) of all the items are enumerated and they are maintained by an item-frequency list together with the maximum error d of a frequency (round numbers). A stream is divided into buckets whose size is equal to ceiling ($1/\varepsilon$). Here ceiling is a roundup function. If an item already exists in a list, its frequency is incremented by one. If an item which should belong to the bucket b is added to the list for the first time, its frequency is initialized to 1 and the error d is set to ($b-1$). If the total number of items reaches twice the bucket size, an item with the frequency $f <= (b-d)$ will be deleted from the list. In this way, the algorithm maintains the size of the item-frequency list so as to fit in memory.

- Trend analysis
Stream data are analyzed by observing changes such as long-term trends, repetitions, seasonal changes, and random changes. Simple methods of detecting changes include a moving average. This

calculates an average or a weighted average, shifting the set of data. The moving average can make streams smooth. In periodic repetitions, discovery of the cycle itself becomes important.

- Similarity search
  There is a task which searches streams similar to a stream given as a sample. It is especially useful in finding a partial match between streams (see Fig. 13.9). However, it is not realistic to compare data which have a large size or large number of dimensions (attributes). That is, it is important to transform the original data into a set of features smaller than the original data. Some methods such as discrete Fourier transformation, discrete wavelet transformation, and principal component analysis can serve that purpose. Although the Euclidean distances can be used as those based on such features in many cases, it is necessary to appropriately identify corresponding parts (e.g., deletion of deficit parts) and adjust different offsets or scales in comparing more than two stream data.

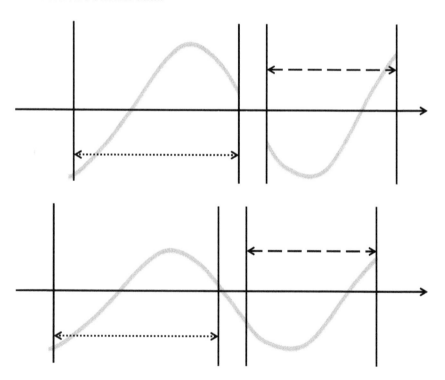

Figure 13.9 Partial matching of streams.

References

[Aboulnaga et al. 2001] A. Aboulnaga, J. Naughton and C. Zhang: Generating Synthetic Complex-Structured XML Data. In Proc. of the ACM SIGMOD Intl. Workshop on the Web and Databases, pp. 79–84 (2001).

[Aggarwal et al. 2003] Charu C. Aggarwal, Jiawei Han, Jianyong Wang and Philip S. Yu: A framework for clustering evolving data streams. In Proc. of the 29th international conference on Very large data base, pp. 81–92 (2003).

[Aggarwal et al. 2007] Charu C. Aggarwal, Na Ta, Na Ta, Jianhua Feng and Mohammed Zaki: XProj: A Framework for Projected Structural Clustering of XML Documents. In Proc. of 13th ACM SIGKDD intl. conf. on Knowledge discovery and data mining (2007).

[Babcock et al. 2002] Brian Babcock, Shivnath Babu, Mayur Datar, Rajeev Motwani and Jennifer Widom: Models and issues in data stream systems. In Proc. of the 21st ACM SIGMOD-SIGACT-SIGART symposium on Principles of database systems (2002).

[Barbosa et al. 2002] D. Barbosa, A. Mendelzon, J. Keenleyside and K. Lyons: ToXgene: an Extensible Template-based Data Generator for XML. In Proc. of the ACM SIGMOD International Workshop on the Web and Databases, pp. 49–54 (2002).

[Chawathe 1999] S.S. Chawathe: Comparing Hierarchical data in external memory. In Proc. of Very Large Data Bases Conference (1999).

[Chen et al. 2003] Yi Chen, Susan Davidson, Carmem Hara and Yifeng Zheng: RRXS: Redundancy reducing XML storage in relations. In Proc. of the 29th VLDB Conference, pp. 189–200 (2003).

[Cohen 2008] S. Cohen: Generating XML Structure Using Examples and Constraints. In Proc. of the Very Large Data Bases Endowment 1(1): 490–501 (2008).

[Dalamagas et al. 2004] T. Dalamagas, T. Cheng, K. Winkel and T. Sellis: Clustering XML Documents Using Structural Summaries. In Proc. of EDBT Workshops on Current Trends in Database Technology (2004).

[Deutsch et al. 1999] Alin Deutsch, Mary Fernandez and Dan Suciu: Storing Semistructured Data with STORED. In Proc. of ACM SIGMOD Conf., pp. 431–442, June 1999.

[Domingos et al. 2000] Pedro Domingos and Geoff Hulten: Mining high-speed data streams, In Proc. of the Sixth ACM SIGKDD International Conference on Knowledge Discovery and Data Mining, pp. 71–80 (2000).

[Florescu et al. 1999] Daniela Florescu and Donald Kossmann: Storing and Querying XML Data Using an RDBMS. IEEE Data Engineering Bulletin 22(3): 27–34 (1999).

[Gaber et al. 2005] Mining Data Streams: A Review: Mohamed Medhat Gaber, Arkady Zaslavsky and Shonali Krishnaswamy. ACM Sigmod Record 34(2) (2005).

[Garofalakis et al. 2000] Minos N. Garofalakis, Aristides Gionis, Rajeev Rastogi, S. Seshadri and Kyuseok Shim: XTRACT: A System for Extracting Document Type Descriptors from XML Documents. In Proc. of ACM SIGMOD Conf., pp. 165–176 (2000).

[Goldman et al. 1997] Roy Goldman and Jennifer Widom: DataGuides: Enabling Query Formulation and Optimization in Semistructured Databases. In Proce. of VLDB Conf., pp. 436–445 (1997).

[Harazaki et al. 2011] Manami Harazaki, Joe Tekli, Shohei Yokoyama, Naoki Fukuta, Richard Chbeir and Ishikawa Hiroshi: XBEGENE: Scalable XML Documents Generator By Example Based on Real Data. In Proc. of Intl. Conf. on Data Engineering and Internet Technology (2011).

[Hulten et al. 2001] G. Hulten, L. Spencer and P. Domingos: Mining time-changing data streams. In Proc. of the seventh ACM SIGKDD international Conference on Knowledge Discovery and Data Mining, pp. 97–106 (2001).

[Inokuchi et al. 2000] A. Inokuchi, T. Washio and H. Motod: An apriori-based algorithm for mining frequent substructures from graph data. In Proc. of Conf. on Principles of Data Mining and Knowledge Discovery, pp. 13–23 (2000).

[Ishikawa et al. 1999] H. Ishikawa, K. Kubota, Y. Noguchi, K. Kato, M. Ono, N. Yoshizawa and Y. Kanemasa: Document warehousing based on a multimedia database system. In Proc. of 15th IEEE International Conference on Data Engineering, pp. 168–173 (1999).

[Ishikawa et al. 2001] Hiroshi Ishikawa, Shohei Yokoyama, Seiji Isshiki and Manabu Ohta: Xanadu: XML- and Active-Database-Unified Approach to Distributed E-Commerce. In Proc. of DEXA Workshop, pp. 833–837 (2001).

[Ishikawa et al. 2007] Hiroshi Ishikawa, Hajime Takekawa and Kaoru Katayama: Proposal and Evaluation of a Technique of Discovering XML Structures for Efficient Retrieval, IADIS International Journal on WWW/Internet 5(1): 80–97 (2007).

[Jiang et al. 2002] Haifeng Jiang, Hongjun Lu, Wei Wang and Jeffrey Xu Yu: Path Materialization Revisited: An Efficient Storage Model for XML Data. In Proc. of Thirteenth Australasian Database Conf., pp. 85–94 (2002).

[Klettke et al. 2001] Meike Klettke and Holger Meyer: XML and Object-Relational Database Systems—Enhancing Structural Mappings Based On Statistics. Lecture Notes in Computer Science, vol. 1997, pp. 151–170 (2001).

[Kuramochi et al. 2001] Michihiro Kuramochi and George Karypis: Frequent Subgraph Discovery. In Proc. of ICDM Conf., pp. 313–320 (2001).

[Liefke et al. 2000] H. Liefke and D. Suciu: XMILL: An efficient compressor for XML data. In Proc. of SIGMOD Conf., pp. 153–164 (2000).

[Manku et al. 2002] Gurmeet Singh Manku and Rajeev Motwani: Approximate Frequency Counts over Data Streams, In Proc. of the 28th International Conference on Very Large Data Base, pp. 346–357 (2002).

[Manning et al. 1999] Christopher D. Manning and Hinrich Schütze: Foundations of Statistical Natural Language Processing. The MIT Press (1999).

[Min et al. 2003] Jun-Ki Min, Myung-Jae Park and Chung Chin-Wan: XPRESS: a queriable compression for XML data. In Proc. of the 2003 ACM SIGMOD intl. conf. on Management of data, pp. 122–133 (2003).

[Moh et al. 2000] Chuang-Hue Moh, Ee-Peng Lim and Wee-Keong Ng: DTD-Miner: A Tool for Mining DTD from XML Documents. In Proc. of Second Intl. Workshop on Advance Issues of E-Commerce and Web-Based Information Systems, pp. 144–151 (2000).

[Shanmugasundaram et al. 1999] Jayavel Shanmugasundaram, Kristin Tufte, Gang He, Chun Zhang, David DeWitt and Jeffrey Naughton: Relational Databases for Querying XML Documents: Limitations and Opportunities. In Proc. of the VLDB Conf., pp. 302–314 (1999).

[Shirai et al. 2013] Motohiro Shirai, Masaharu Hirota, Hiroshi Ishikawa and Shohei Yokoyama: A method of area of interest and shooting spot detection using geo-tagged photographs, In Proc. of ACM SIGSPATIAL Workshop on Computational Models of Place 2013 at ACM SIGSPATIAL GIS 2013 (2013).

[Tan et al. 2002] Pang-Ning Tan and Vipin Kumar: Discovery of Web Robot Sessions based on their Navigational Patterns. Data Mining and Knowledge Discovery 6(1): 9–35 (2002).

[Tolani et al. 2002] Pankaj M. Tolani and Jayant R. Haritsa: XGRIND: A Query-Friendly XML Compressor. In Proc. of IEEE ICDE Conf., pp. 225–234 (2002).

[van Leuken et al. 2009] Reinier H. van Leuken, Lluis Garcia Pueyo, Ximena Olivares and Roelof van Zwol: Visual diversification of image search results. In Proc. of WWW, pp. 341–350 (2009).

[Wu et al. 2001] Xiaoying Wu, Tok Wang Ling, Sin Yeung Lee, Mong-Li Lee and Gillian Dobbie: NF-SS: A Normal Form for Semistructured Schema. In Proc. of ER Workshops, pp. 292–305 (2001).

[XML Schema 2014] XML Schema http://www.w3.org/XML/Schema Accessed 2014

[XQuery 2014] XQuery http://www.w3.org/TR/xquery/ Accessed 2014

[Yoshikawa et al. 2001] M. Yoshikawa and T. Amagasa: XRel: A path-based approach to storage and retrieval of XML documents using relational databases. ACM Transactions on Internet Technology 1(1): 110–141 (2001).

[Zaki 2002] Mohammed Javeed Zaki: Efficiently mining frequent trees in a forest. In Proc. of KDD Conf., pp. 71–80 (2002).

[Zaki et al. 2003] M.J. Zaki and C.C. Aggarwal: XRules: an effective structural classifier for XML data. In Proc. of the ninth ACM SIGKDD intl. conf. on Knowledge discovery and data mining (2003).

[Zloof 1977] M. Zloof: Query By Example. IBM Systems Journal 16(4): 324–343 (1977).

Scalability and Outlier Detection

In this chapter, several attempts related to scalability for mining tasks, such as association analysis, clustering, and classification, which target social big data will be introduced. In addition, outlier detection will be described.

14.1 Scalability Related to Association Analysis

Generally, parallel versions of basic techniques are present as countermeasures to the scalability [Zaki 1999]. Parallel technologies can be divided into two categories based on whether or not to share memories among processors. Methods which share memories, called shared-memory, in general, can directly access all the memories of the system in a unified manner. Therefore programming is simplified, but scalability as to the processors is limited by the bandwidth of the bus for data transfer. On the other hand, methods where each processor has a memory of its own and does not share memory with others, called shared-nothing, access the memories of the other processors by way of messages sent to them as needed. Therefore, simplicity of programming is sacrificed, but the problem of scalability can be solved in a straightforward manner.

14.1.1 Shared-nothing

First, methods that do not share memory will be described.

(1) Count distribution based on Apriori

In this method, each processor has a local database that can be constructed by partitioning the global database [Agrawal et al. 1996]. A hash tree is made so as to count candidate global frequent itemsets C_k from global frequent itemsets L_{k-1}. Each processor counts supports based on its local database, exchanges the result with the others, and obtains global support counts. Each processor produces C_k in parallel from thus obtained L_{k-1}. This process is repeated until all frequent itemsets are found.

(2) DHP-based method

This method has hash tables so as to count local supports for 1-itemsets and approximate supports for 2-itemsets based DHP (Dynamic Hashing and Pruning) [Park et al. 1995]. By all-to-all broadcasting, each processor can obtain global support counts of the 1-itemset. For the 2-itemsets, only entries in the hash tables which are found frequently are exchanged. Hash tables are used so as to obtain the global support counts for 2-itemsets. Hereafter, C_k are produced from L_{k-1} (k > 2) similarly to the Apriori algorithm.

(3) P2P partitioning-based method

The work of the author and his colleagues [Ishikawa et al. 2004] was intended to realize parallel mining at large-scale and low-cost by the P2P (Peer to Peer) network. This method can be classified into one based on both partitioning and shared-nothing. It partitions and exchanges the local itemsets between the locally connected processors on the basis of the protocol P2P. Thereby the control can be simplified and the network traffic can be reduced.

In order to perform load balancing, a rank is considered as an indicator of the processing power of each node. The rank of the node is determined from the ranks of the nodes adjacent to the former node by considering the influence of the transmission delay. By using a distributed hash table, the proposed method makes each node have the information about the processing capabilities of its adjacent nodes. Thereby the method can compute the ranks without the need for a central server and perform load balancing.

14.1.2 Shared-memory

Here the shared memory approach will be explained.

(1) Zaki's Apriori-based method

In this method [Zaki et al. 1997], each processor has its own partition of the same size that is made by logically dividing the whole data and shares a hash tree for candidate global frequent itemsets. The locking mechanism in units of each leaf node is used for simultaneous updating of the hash tree. Each processor computes the support counts of frequent itemsets on the basis of its logical partition.

(2) DIC-based method

This method [Cheung et al. 1996] assigns *l* virtual partitions to *p* processors assuming that $l \geq p$. Letting *m* be the size of the item set, *l* vectors in *m*-dimensional space are considered so as to compute the local counts for each partition. Based on the distance between the vectors, clustering is performed so as to maximize the inter-cluster distance and to minimize the intra-cluster distance. Let *k* be the number of clusters here. The same number of elements are selected from each of the *k* clusters and assigned to each processor because DIC requires homogeneous partitions. Further, so as to make *r* homogeneous subpartitions within each processor, *k* clusters are created and the same number of elements are selected from each of the *k* clusters and assigned to the subpartitions.

14.2 Approach to Scalability for Clustering

Approaches to scalability for clustering include the following.

- Introduce a hierarchical structure and aggregate data based on the structure.
- Use random sampling.
- Partition data and use the data density.

The following sections will describe hierarchical approaches, density-based approaches, and graph mining as well as approaches to detecting outliers in relation to clustering.

14.2.1 Hierarchical Approaches

BIRCH [Zhang et al. 1996], CURE [Guha et al. 1998], and Chameleon [Karypis et al. 1999] are among typical variations of hierarchical clustering algorithms. Here CURE and BIRCH will be described while Chameleon will be explained as a clustering method based on graphs.

(1) BIRCH

BIRCH introduces the concept of a CF tree. BIRCH performs hierarchical clustering using the CF tree and performs a refinement of the constructed clusters. The CF tree is a balanced tree where each node represents a cluster. The nodes store *CF* (clustering feature) values that represent features of child nodes, that is, subclusters (see Fig. 14.1).

- *CF* value = $(m, \mathbf{LS}, \mathbf{SS})$

The subcluster includes *m* multi-dimensional vectors d_i ($i = 1$, *m*). *LS* and *SS* are the sum of the vectors and the sum of the squared vectors, respectively. The average (i.e., the centroid of the cluster)

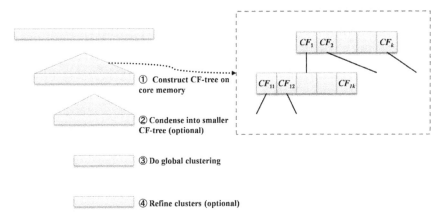

Figure 14.1 BIRCH.

and the variance can be incrementally calculated by using LS and SS. These values are also used in calculating the distance between clusters. The nonleaf nodes of the CF tree have more than one child and the number B of such children is determined by parameters such as the size of pages and the number of attributes. The nonleaf node stores the CF value and pointer for each child. The leaf nodes (clusters) also have multiple entries, where each entry represents the number of scanned vectors. The steps of clustering based on CF tree are illustrated in Fig. 14.1.

(2) CURE

CURE aims to form non-spherical clusters using representative objects (see Fig. 14.2). CURE has increased scalability with respect to the algorithm, based on random sampling. CURE determines the size of sample data in random sampling so as to obtain at least a certain number of objects from each cluster.

(Algorithm)
1. Draw random samples from N pieces of data;
2. Make P partitions out of the random samples;
3. Cluster each partition around the N/PR representative points; N/R subclusters can be obtained as a result; Here R is the desired reduction of data in a partition;
4. Apply the hierarchical agglomerative clustering to such subclusters from bottom upto the desired number of clusters (for example, as two clusters connected by the dashed line in Fig. 14.2a);
5. Mark each piece of remaining data with the label of the nearest cluster.

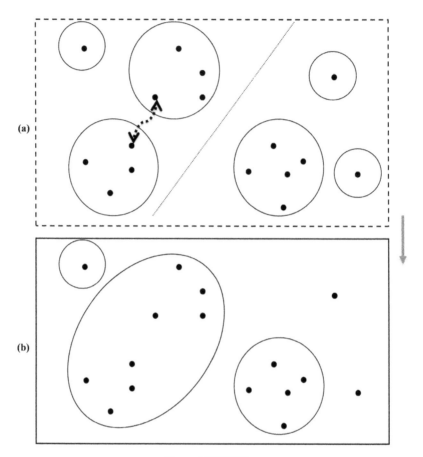

Figure 14.2 CURE.

14.2.2 Density-based Clustering

In this section, the clustering algorithms based on the concept of spatial density with physical analogy will be introduced. Density-based clustering can be divided into the centroid-based approaches and the grid-based approaches. Approaches based on division of the low-dimensional space containing data open up the potential for improved scalability by parallel processing.

(1) Centroid-based approach

First, DBSCAN [Ester et al. 1996] will be explained as an approach based on the concept of traditional density. DBSCAN performs clustering by focusing on the density based on the centroid of the cluster. Let *MinP* and ε be the minimum number of objects within a cluster and the radius of the cluster, respectively. DBSCAN can cluster spatial objects of arbitrary shape within the maximal range as to the following density-connectivity.

(Definition) Density-connectivity
- The range of ε-radius centered on an object is called ε-neighborhood.
- The ε-neighborhood of an object that contains at least *MinP* (i.e., a threshold parameter) objects is called a core object.
- An object that is within the ε-neighborhood of a core object but is not a core object itself is called a border object.
- An object that is neither a core object nor a border object also is called a noise object.
- If an object p is in the ε-neighborhood of an object q, p is said to be directly density-reachable from q.
- If p_{i+1} is directly density-reachable from p_i ($p_1 = q$ and $p_n = p$) in a sequence $\{p_1 p_2 \dots p_n\}$, p is said to be indirectly density-reachable, shortly, density-reachable from q.
- If p and q are density-reachable from another object, q and p are said to be density-connected.

The DBSCAN algorithm performs clustering according to the above definitions as follows.

(Algorithm) DBSCAN
1. Classify objects into core, border, or noise objects.
2. Remove the noise objects.
3. Connect core objects in the ε-neighborhood of each other with an edge.
4. Group thus connected core objects into a separate cluster.
5. Assign border objects to the cluster to which the associated core objects belong.

Density reachability constitutes a transitive closure of direct density reachability. Therefore clusters is the maximal set of objects that are density-connected to each other. For example, let the radius of the circles and *MinP* be ε and 3, respectively, P_2 and P_1 will be core objects, but *P3* will be a border object (see Fig. 14.3). However, P_3 can be density-reachable via P_2 from P_1, thus P_1, P_2, and P_3 are density-connected (see Fig. 14.3a). Similarly, O, Q, and R are all density-connected, all these objects will belong to the same cluster as a result (see Fig. 14.3b). Objects that are not included in the cluster can be eliminated by this algorithm since they are noise objects (i.e., outliers).

The time complexity of DBSCAN is primarily N^* (time required to search for points in the ε-neighborhood) if N is the number of objects. Indeed, this is $O(N^2)$ at worst, but use of the hierarchical indexes such as R*tree, and kd-tree in the low-dimensional space can lower the complexity to $O(N \log N)$.

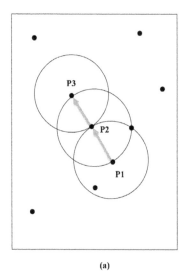 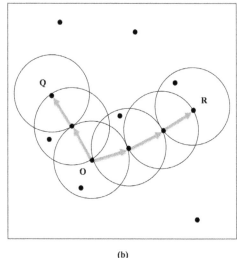

(a)   (b)

Figure 14.3 DBSCAN.

DBSCAN can find clusters with the desirable shapes and sizes and be relatively robust to noises unlike *k*-means. Further, the number of clusters is automatically determined. DBSCAN has also problems; It is rather difficult to define meaningful concepts of density in the case of the high-dimensional space and the algorithm doesn't work well in case clusters have non-uniform densities.

Anyhow, DBSCAN is suitable for clustering a set of images with the position information on the maps as described in the previous section.

(2) Grid-based clustering

Let us consider a multidimensional space where each dimension corresponds to a single attribute. It is assumed that each dimension consists of a series of adjacent intervals. The algorithm clusters a set of points contained in the grid cells, or simply cells, surrounded by intervals in multi dimensions. It is assumed that in advance all data are associated with the cells and information such as the number of data contained in the cell is aggregated.

The width of the interval is usually determined by performing binning, which is used in the discretization of continuous values. Therefore, intervals in each dimension have properties such as constant width and constant frequency, according to the used methods as the binning.

DENCLUE [Hinneburg et al. 1998] will be described as an algorithm for clustering based on grid concepts as follows. DENCLUE (DENsity CLUstEring) models the density function of a set of points as the sum of influence functions of each point. Data are collected around the points of high density. The following influence function is commonly used.

(Influence function) It is a symmetric function of the influence of the point y on the point x.

- $$f_{Gauss}(x, y) = e^{-\frac{d(x,y)^2}{2\sigma^2}}$$

Here D is the distance between x and y and σ is the parameter that controls the attenuation of the influence.

The density function of a point is the sum of the influence functions originating from all the other points. In general, the density function of the point set has local peaks called local density attractors.

The outline of the algorithm is follows.

For each point, find a local density attractor by using the hill-climbing method [Russell et al. 2003] and assign the point to the density attractor, thereby making a cluster. Points associated with the density attractor whose peak is below the prescribed threshold ξ are removed as noises. If the densities of all the points that connect two peaks as illustrated by the solid line in Fig. 14.4 are higher than or equal to ξ, clusters associated with these peaks are merged into one.

Since the density of one point is the sum of the densities of all other points, the computational cost basically becomes $O(N^2)$, letting N be the number of data points. Therefore, DENCLUE addresses this problem by dividing a region containing all data points into the grid of cells. Cells that contain no points are not considered. The cells and associated information are accessed by using a tree-structured index. So as to calculate the density of a point and find the density attractor closest to the point, only the cell that contains the point and cells that can be connected with the former cell are searched. Although the accuracy on the estimate of the density may be decreased due to this scheme, it is possible to reduce the computational cost.

Density

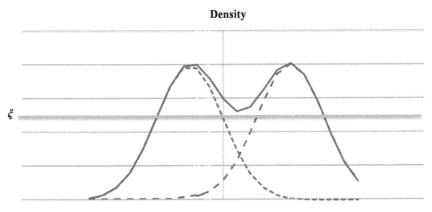

Figure 14.4 DENCLUE.

As DENCLUE is based on the density function, DENCLUE can make more accurate and flexible estimates of the density than DBSCAN. Similarly to the other methods, however, DENCLUE does not work well in some cases such as the non-uniformity of density and high-dimensional data. Please note that the parameters σ and ξ may greatly affect the final clusters.

14.2.3 Graph Clustering

First, let data and proximities between the data be the nodes and weighted edges of the graph, respectively. Chameleon, the method of performing clustering based on the graph will be described.

Chameleon [Karypis et al. 1999] is a kind of hierarchical clustering method for graphs. If one of two nodes is among the k-nearest neighbors of the other, those two nodes are connected with the edge whose weight represents similarity between the nodes. The thus constructed graphs are called k-nearest neighbor graphs. The graphs (b), (c) and (d) of Fig. 14.5 are 1-,2-, and 3-nearest neighbor graphs, respectively, for the graph (a).

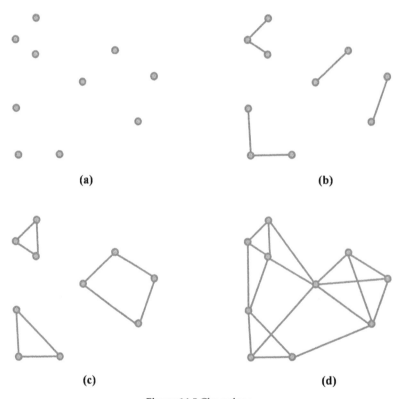

Figure 14.5 Chameleon.

The outline of the Chameleon algorithm will be described below.

(Algorithm) Chameleon
1. Construct k-nearest neighbor graphs based on the k-nearest neighbors of all data.
2. Partition the graphs into subgraphs (i.e., clusters) whose sizes are less than or equal to the threshold parameter.
3. Repeat the following step until no clusters remain to be merged{// agglomeration method.
4. Choose and merge the clusters whose self-similarity based on the combination of RC (relative closeness) and RI (relative interconnectivity) is preserved best.}

Here RC and RI, which will be defined below, are used in the measure for cluster self-similarity, that is, similarity between clusters to be merged and the merged cluster.

(Definition) relative closeness (RC)

- $$RC = \frac{S(C_i,C_j)}{\frac{m_i}{m_i+m_j}S(C_i)+\frac{m_j}{m_i+m_j}S(C_j)}$$

Here m_i and m_j are the sizes of the clusters C_i and C_j, respectively. $S(C_i, C_j)$ is the average of the weights of the edges connecting C_i and C_j (i.e., k-nearest neighbors). $S(C_i)$ is the average of the weights of the edges in the case where C_i is divided into two clusters. $S(C_j)$ is defined similarly.

(Definition) relative interconnectivity (RI)

- $$RI = \frac{E(C_i,C_j)}{\frac{1}{2}(E(C_i)+E(C_j))}$$

Here $E(C_i, C_j)$ is the sum of the weights of the edges connecting C_i and C_j (i.e., k-nearest neighbors). $E(C_i)$ is the sum of the weights of the edges in the case where C_i is divided into two clusters. $E(C_j)$ is defined similarly.

As cluster self-similarity, the combination, for example, "$RC(C_i, C_j)^{\alpha} * RI(C_i, C_j)$" are used. Here α is a parameter for the user to specify, which will be usually greater than 1.

K-nearest neighbor graphs can be constructed at the computational cost of $O(N\log N)$. So as to partition graphs into two, Chameleon uses the program called hMETIS, whose computational cost is $O(N)$, assuming that N is the number of data points. Therefore the cost of $O(N\log(N/P))$ is required to split into P pieces of graphs. Hierarchical agglomerative clustering can be carried out at the computational cost of $O(N^2\log N)$. Since it is necessary to bisect $(P-i)$ subclusters for the calculation of the similarity at the i-th iteration in clustering, the cost of $O(PN)$ is required. So the computational cost as a whole is $O(PN + N\log N + N^2\log N)$.

Chameleon is relatively robust to the existence of outliers and noises due to its use of the *k*-nearest neighbors. Chameleon, however, assumes that set of points created by partitioning correspond to clusters. Therefore, it has a weak point in that errors remain to be corrected if partitioning itself is wrong.

14.3 Scalability for Classification and Other Tasks

Since basic classification algorithms assume that the whole training set is included in memory, they may cause problems in performance if the size of the training set becomes very large. On the other hand, it is possible to construct classifiers by the sampling or partitioning of the training set similar to association rule mining, but such approaches could not reach the accuracy of the classifier learned using all the data simultaneously.

One of approaches to scalability for the algorithms to produce a decision tree is to devise special data structures. The systems extended in this direction include SLIQ [Mehta et al. 1996] and SPRINT [Shafer et al. 1996] as well as RainForest [Gehrke et al. 1998] and Boat [Gehrke et al. 1999] as the latter's improved versions.

SLIQ has only partial information of the data in memory. First, an identifier (*RID*) is given to each record. One record is represented by separate tables, called attribute lists, which correspond to attributes (see Fig. 14.6). Each attribute list consists of *RID* and corresponding attribute values. The attribute lists are sorted by the attribute values. Further, another table, called class list, which consists of *RID* and the class name is made. Based the above data structures, with the *RID* as index key, the class names and attribute values of records can be accessed. Basically, while the attribute lists are stored on disks, the class list resides in memory. The amount of the memory to be used at a time is proportional to the size of the class list (i.e., the size of the training set).

SPRINT, RainForest, and Boat, commonly represent one record by providing each attribute with a list which consists of *RID*, attribute value, and class name. Each attribute list is also ordered according to the attribute values. Tables for attribute lists are usually distributed with the order of the

RID	Travel
1	dislike
2	dislike
4	dislike
3	like

RID	Age
2	23
3	35
1	38
4	47

RID	Purchase wine	LeafID
1	yes	
2	yes	
3	no	
4	no	

Figure 14.6 SLIQ.

records reserved according to the partitioning of the data into the proper locations. Examples by SPRINT are illustrated in Fig. 14.7.

Another approach to the scalability of classification is parallelization. Ensemble learning constructs one classifier with high accuracy with the use of multiple separate classifiers. It is suitable for performing classification tasks in a parallel fashion regardless of the type of classifiers. So ensemble-learning an individual classifier (e.g., decision tree) in parallel can be considered as promising. For SVM, in particular, there is an attempt to perform the parallelization by Cascade process [Graf et al. 2004]. There exist similar attempts based on distributed computing in P2P environments for decision tree learning [Bhaduri et al. 2008a] and multivariate analysis [Bhaduri et al. 2008b]. Others attempt to perform parallel processing by using the multi-core computers for data mining tasks including the Naive Bayes [Chu et al. 2006].

Last in this section, the scalability to the power methods, which are frequently used in the Web structure mining will be briefly described. In order to compute PageRank, an incremental approach is used [Desikan et al. 2005]. First, this method calculates the PageRanks for all pages once. After that, PageRanks are recalculated only for pages that have been changed with respect to the link structures and for pages that will be affected by the former pages on the Web. Then the results are integrated with PageRanks of the remaining pages. Another method [Gleich et al. 2004] divides the matrix representing the Web graphs so as to calculate the ranks in parallel. A third method calculates the approximate eigenvector by using the Monte Carlo method in a small number of iterations [Avrachenkov et al. 2007].

Travel	Purchase wine	RID		Age	Purchase wine	RID
dislike	yes	1		23	yes	2
dislike	yes	2		35	yes	3
dislike	no	4		38	no	1
like	no	3		47	no	4

Figure 14.7 SPRINT.

14.4 Outlier Detection

So as to ensure scalability for big data processing, random sampling is often useful. However, random sampling is not applicable if the objective is to discover a phenomenon that does not occur so frequently. For example, in the detection of the Higgs boson that are produced only once in 10 trillion times of particle collision in theory, it is necessary to exhaustively analyze all the experimental results.

In general, outlier detection is to detect objects with values different from those of many other objects. For example, applications of outlier detection include the following.

- Detection of frauds
- Detection of system intrusion
- Prediction of an abnormal phenomenon
- Detection of side effects of vaccination in public health
- Detection of side effects of drugs in medicine

Typical approaches to outlier detection include the following.

- model-based approach
- proximity-based approach
- density-based approach
- clustering-based approach

In the first place, what are outliers? According to D.M. Hawkins [Hawkins 1980], the definition of outliers is as follows.

(Definition) Hawkins' outlier
Outliers are observed data which are very different from other observed data as if they were caused by a mechanism totally different from the usual one.

This abstract definition, however, doesn't tell us how to find outliers. The above mentioned approaches provide the appropriate definitions of outliers. Let's check them below.

(1) Model-based approach

The approach based on statistical models assumes that an outlier is an object with a low probability in the probability distribution. For example, objects following the normal distribution are regarded as outliers if their distances from the average exceed the threshold value.

(2) Proximity-based approach

This approach determines outliers by using their distances to the k-nearest neighbors. For example, the minimum radius of the k-nearest neighbors of an object is regarded as the degree of outlier-ness of the object. Thus, the larger the minimum radius of the object, the higher the degree that the object is an outlier.

(3) Density-based approach

This approach determines whether an object is an outlier by the inverse of the density around the object. The degree of outlier-ness is higher if the value is greater. For example, as the density of data, the number of objects within a specified distance and the inverse of the average of the distances between the object and the k-nearest neighbors are used.

(4) Clustering-based approach

This approach determines that an object is an outlier if the object does not *strongly* belong to any cluster. For example, an object whose distance from the centroid of the cluster to which it belongs in divisive clustering is greater than the threshold and another which is lastly merged in hierarchical agglomerative clustering are considered as outliers. In other cases, objects belonging to small-sized clusters may simply be regarded as outliers.

References

[Agrawal et al. 1996] R. Agrawal and J. Schafer: Parallel Mining of Association Rules. IEEE Transactions on Knowledge and Data Engineering 8(6): 962–969 (1996).

[Avrachenkov et al. 2007] K. Avrachenkov, N. Litvak, D. Nemirovsky and N. Osipova: Monte Carlo Methods in PageRank Computation: When One Iteration is Sufficient. SIAM J. Numer. Anal. 45(2): 890–904 (2007).

[Bhaduri et al. 2008a] K. Bhaduri and H. Kargupta: An Efficient Local Algorithm for Distributed Multivariate Regression in Peer-to-Peer Networks. SIAM International Conference on Data Mining, Atlanta, Georgia, pp. 153–164 (2008).

[Bhaduri et al. 2008b] K. Bhaduri, R. Wolff, C. Giannella and H. Kargupta: Distributed Decision Tree Induction in Peer-to-Peer Systems. Statistical Analysis and Data Mining 1(2): 85–103 (2008).

[Cheung et al. 1996] D.W. Cheung, J. Han, V. Ng, A. Fu and Y. Fu: A fast distributed algorithm for mining association rules. In Proc. of Int. Conf. Parallel and Distributed Information Systems, pp. 31–44 (1996).

[Chu et al. 2006] C.T. Chu, S.K. Kim, Y.A. Lin, Y. Yu, G. Bradski, A.Y. Ng and K. Olukotun: Map-reduce for machine learning on multicore. In NIPS 6: 281–288 (2006).

[Desikan et al. 2005] Prasanna Desikan, Nishith Pathak, Jaideep Srivastava and Vipin Kumar: Incremental page rank computation on evolving graphs. In Special Interest Tracks and Posters of the 14th International Conference on World Wide Web (WWW '05). ACM (2005).

[Ester et al. 1996] Martin Ester, Hans-Peter Kriegel, Jörg Sander and Xiaowei Xu: A density-based algorithm for discovering clusters in large spatial databases with noise. In Proc. of the Second Intl. Conf. on Knowledge Discovery and Data Mining, pp. 226–231 (1996).

[Gehrke et al. 1998] Johannes Gehrke, Raghu Ramakrishnan and Venkatesh Ganti: RainForest —A Framework for Fast Decision Tree Construction of Large Datasets. In Proceedings of the 24rd International Conference on Very Large Data Bases (VLDB '98), pp. 416–427 (1998).

[Gehrke et al. 1999] Johannes Gehrke, Venkatesh Ganti, Raghu Ramakrishnan and Wei-Yin Loh: BOAT—optimistic decision tree construction. ACM SIGMOD Rec. 28(2): 169–180 (1999).

[Gleich et al. 2004] David Gleich, Leonid Zhukov and Pavel Berkhin: Fast parallel PageRank: A linear system approach. Yahoo! Research Technical Report YRL-2004-038 (2004).

[Graf et al. 2004] Hans Peter Graf et al.: Parallel Support Vector Machines: The Cascade SVM. In NIPS. 2004, pp. 521–528 (2004).

[Guha et al. 1998] Sudipto Guha, Rajeev Rastogi and Kyuseok Shim: CURE: An Efficient Clustering Algorithm for Large Databases. In Proc. of the ACM SIGMOD intl. conf. on Management of Data, pp. 73–84 (1998).

[Hawkins 1980] D. Hawkins: Identification of Outliers. Chapman and Hall (1980).

[Hinneburg et al. 1998] Alexander Hinneburg and Daniel A. Keim: An Efficient Approach to Clustering in Large Multimedia Databases with Noise. In Proc. of Intl. Conf. on Knowledge Discovery and Data Mining, pp. 58–65 (1998).

[Ishikawa et al. 2004] Hiroshi Ishikawa, Yasuo Shioya, Takeshi Omi, Manabu Ohta and Kaoru Katayama: A Peer-to-Peer Approach to Parallel Association Rule Mining, Proc. 8th

International Conference on Knowledge-Based Intelligent Information and Engineering Systems (KES 2004), pp. 178–188 (2004).

[Karypis et al. 1999] George Karypis, Eui-Hong Han and Vipin Kumar: CHAMELEON: A Hierarchical Clustering Algorithm Using Dynamic Modeling. IEEE Computer 32(8): 68–75 (1999).

[Mehta et al. 1996] Manish Mehta, Rakesh Agrawal and Jorma Rissanen: SLIQ: A Fast Scalable Classifier for Data Mining. In Proceedings of the 5th International Conference on Extending Database Technology: Advances in Database Technology (EDBT '96), Springer-Verlag, pp. 18–32 (1996).

[Park et al. 1995] Jong Soo Park, Ming-Syan Chen and Philip S. Yu: An Effective Hash-Based Algorithm for Mining Association Rules. In Proc. of the 1995 ACM SIGMOD Intl. Conf. on Management of Data, pp. 175–186 (1995).

[Russell et al. 2003] Stuart Jonathan Russell and Peter Norvig: Artificial Intelligence: A Modern Approach. Pearson Education (2003).

[Shafer et al. 1996] John C. Shafer, Rakesh Agrawal and Manish Mehta: SPRINT: A Scalable Parallel Classifier for Data Mining. In Proceedings of the 22nd International Conference on Very Large Data Bases (VLDB '96), Morgan Kaufmann Publishers Inc., pp. 544–555 (1996).

[Zaki et al. 1997] M. Zaki, S. Parthasarathy, M. Ogihara and W. Li: New Algorithms for Fast Discovery of Association Rules. In Proc. of 3rd ACM SIGKDD Int. Conf. on Knowledge Discovery and Data Mining, pp. 283–296 (1997).

[Zaki 1999] M.J. Zaki: Parallel and Distributed Association Mining: A Survey. IEEE Concurrency 7(4): 14–25 (1999).

[Zhang et al. 1996] Tian Zhang, Raghu Ramakrishnan and Miron Livny: BIRCH: an efficient data clustering method for very large databases. In Proc. of the ACM SIGMOD intl. Conf. on Management of Data, pp. 103–114 (1996).

Appendix I

Capabilities and Expertise Required for Data Scientists in the Age of Big Data

Data analysts are also called data scientists. In the era of big data, data scientists are in more and more demand. At the end of this part, capabilities and expertise necessary for big data scientists will be summarized. They include at least the following items. (Please note that this book explains the topics relevant to the underlined items in detail in separate chapters.)

- Can construct a hypothesis
- Can verify a hypothesis
- Can mine social data as well as generic Web data
- Can process natural language information
- Can represent data and knowledge appropriately
- Can visualize data and results appropriately
- Can use GIS (geographical information systems)
- Know about a wide variety of applications
- Know about scalability
- Know and follow ethics and laws about privacy and security
- Can use security systems
- Can communicate with customers

According to the order of the above items, supplementary explanations will be made.

Needless to cite as a successful example in data-intensive scientific discovery, that hypotheses on the Higgs boson have been confirmed with high probability by independent analyses of a tremendous amount of experimental data, the role of hypotheses has been more important in the age of big data than ever. As repeatedly mentioned, construction of hypotheses prior to analysis is mandatory for appropriate collection of data, appropriate selection of already collected data, and appropriate adoption of confirmed hypotheses as analytic results. As data mining is helpful for hypothesis construction, applicable knowledge about such technologies is necessary. In order for hypotheses as analytical results to widely be accepted, they must be quantitatively confirmed. So applicable knowledge about statistics and multivariate analysis is also required.

Generally, both physical real world data without explicit semantics and social data with explicit semantics constitute big data. Integrated analysis of both kinds of data is considered to become more necessary in most of big data applications. As social data are basically on the Web, applicable knowledge about Web mining is necessary. Moreover, since social data are usually described in natural languages, working knowledge of natural language processing for analyzing social data, in particular, text mining is desirable.

Since representation of knowledge constructed from hypotheses or intermediate data corresponding to them significantly determines whether subsequent tasks will be smoothly processed by computers, appropriate representation of data is strongly desirable. So practical knowledge about data and knowledge representation is necessary. Similarly, intermediate and final results of analysis need to be understandably summarized for data scientists and domain experts. Appropriate visualization of analytic summaries enables them to understand constructed hypotheses, discover new insights, and construct further hypotheses. Applicable knowledge about visualization tools is also desirable.

Nowadays, geographical and temporal information are added to collected data in many applications. In such cases, geographical information systems (GIS), which are based on mapping, can be used as visualization platforms. In particular, registration of Mt. Fuji as a world heritage site and selection of Tokyo as the 2020 Olympic venue will propel tourism sectors in Japan to develop big data applications associated with GIS from now on. In such cases, applicable knowledge about GIS which are also aware of temporal information is helpful. Data scientists should be interested in or knowledgeable about both a wide variety of application domains and people involved in such domains.

Scalable systems or tools can process a larger amount of data in practical time if more processing power can be provided in some way (e.g., scale-up and scale-out). Data scientists must be able to judge whether available systems or tools are scalable. In particular, knowledge about scale-out by parallel and distributed computing, which are currently mainstream technologies in Web services, is desirable.

As is not limited to social data, data generated by individuals can be used only if their concerns about invasion of privacy are removed. So as to protect user privacy, both service providers and users are required to respect relevant ethics and policies and follow relevant laws and regulations. However, it is also true that there do exist some people who neglect them and commit crimes. So it is necessary to know about security as a mechanism for protecting data and systems as well as user privacy from such harms.

Last but not least, construction of promising hypotheses requires communication capabilities to extract interests and empirical knowledge as hints from domain experts, formulate hypotheses based on such interests and knowledge, and explain formulated hypotheses and analysis results to the domain experts in appropriate terms.

As the readers have already noticed, it is very difficult for a single data scientist to have all the above capabilities at high levels. In other words, not a single super data scientist but a team of capable persons must be in charge of analyzing and utilizing big data. Thus, constructing big data applications, aimed for discovering collective knowledge or wisdom, requires a team which has diversity as to capabilities among members.

Generally speaking, it is very seldom that a sufficient amount of information and knowledge are provided in advance. Therefore, if one more capability could be added to the above list, it would be fertile imagination.

Appendix II

Remarks on Relationships Among Structure-, Content-, and Access Log Mining Techniques

So far, structure mining, content mining, and access log mining techniques have been described as separate techniques, whether they are targeted at Web data, XML data, or social data. However, they are related to each other.

Needless to say, basic mining techniques such as association analysis, cluster analysis, and classification can be applied to any of structure-, content-, and access log mining. As described in the previous chapter, if access log data are represented as tree structures, then problems in access log mining can be translated into those in structure mining.

Below, content mining and structure mining will be picked up and concretely discussed from this viewpoint.

The first step in analyzing the contents of tweets is to find frequent terms and frequent co-occurring terms. The second step is to correspond terms and co-occurrence relationships between terms to nodes of graphs and edges between nodes, respectively. Please note that only terms and co-occurrences with the frequencies above the specified thresholds are usually included for elements of graphs for practical reasons. The next step is to find terms corresponding to nodes with high centrality such as betweenness centrality, which is defined as follows.

(Definition) Betweenness centrality

Betweenness centrality of a node is the total number of the shortest paths between the other two nodes passing through the node divided by the total number of all the shortest paths between those two nodes.

The other centralities include degree centrality based on the degrees of a node and closeness centrality based on the inverse of the sum of all the shortest distances between a node and every other node. Anyhow, the above approach can be considered a solution by translating problems in content mining into those in structure mining.

For example, with the help of *spurious* correlations, reasons for rapid increase in the number of passengers (i.e., physical real world data) riding from a specific station during a specific period can be found by filtering a set of tweets (i.e., social data) posted around the station during the same period and focusing on the terms within the set which correspond to nodes with such high centrality as described above.

Index

Color Plate Section

Chapter 1

Figure 1.1 Twitter.

Figure 1.2 Flickr.

Figure 1.3 YouTube.

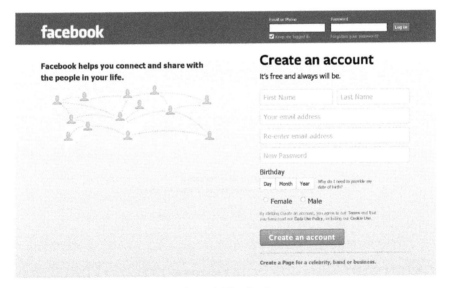

Figure 1.4 Facebook.

Chapter 2

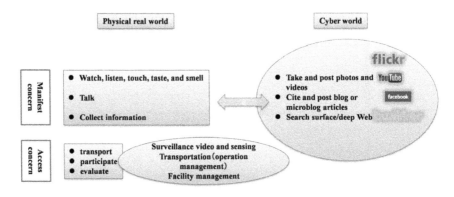

Figure 2.2 Physical real world data and social data.

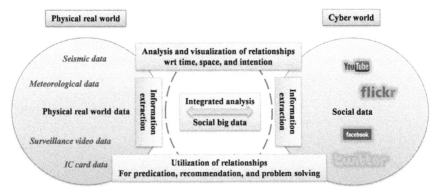

Figure 2.3 Integrated analysis of physical real world data and social data.

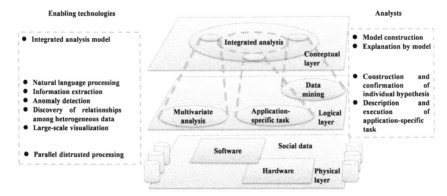

Figure 2.4 Reference architecture for social big data.

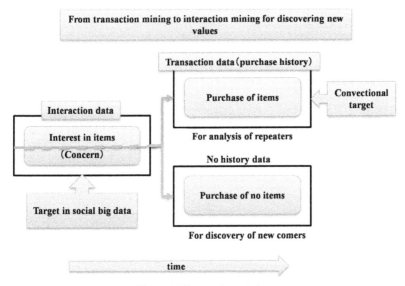

Figure 2.8 Interaction mining.

Chapter 12

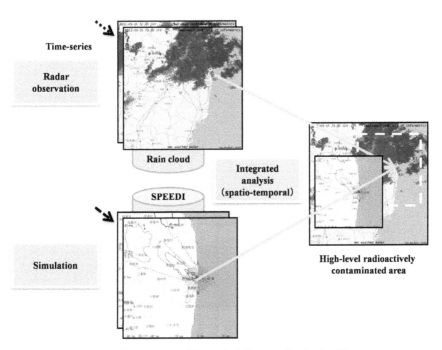

Figure 12.8 Construction of inter-disciplinary collective intelligence.

Chapter 13

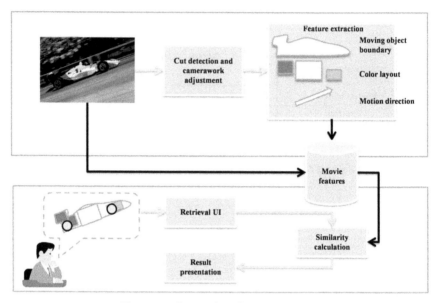

Figure 13.8 Content-based movie retrieval.

Milton Keynes UK
Ingram Content Group UK Ltd.
UKHW031130141024
449569UK00006B/312